Bright Galaxies, Dark Matter, and Beyond

Bright Galaxies, Dark Matter, and Beyond

The Life of Astronomer Vera Rubin

Ashley Jean Yeager

The MIT Press
Cambridge, Massachusetts
London, England

This book was set in Stone Serif and Stone Sans by Westchester Publishing Services. Printed and bound in the United States of America.

Library of Congress Cataloging-in-Publication Data

Names: Yeager, Ashley Jean, author.
Title: Bright galaxies, dark matter, and beyond : the life of
    astronomer Vera Rubin / Ashley Jean Yeager.
Description: Cambridge, Massachusetts : The MIT Press, [2021] |
    Includes bibliographical references and index.
Identifiers: LCCN 2020047103 | ISBN 9780262046121 (hardcover)
Subjects: LCSH: Rubin, Vera C., 1928–2016. | Astronomers—United
    States—Biography. | Women astronomers—United States—
    Biography. | Dark matter (Astronomy)
Classification: LCC QB36.R83 Y43 2021 | DDC 520.92 [B]—dc23
LC record available at https://lccn.loc.gov/2020047103

10 9 8 7 6 5 4 3 2 1

For D'Arcy, Ella, and all young girls
who dream of becoming scientists

Contents

Prologue

A flip of a switch, and astronomer Vera Rubin disappeared. The darkness of the telescope dome swallowed her hourglass frame as she quickly and confidently took a few steps, grabbed the staircase banister, and climbed up. At the top of the stairs, she slid her hand across the door, found the knob, and pushed. Nothing happened. Like a football lineman, she lowered her center of gravity and threw her weight against the hinged hunk of metal, bumping it open with her hip.

A cool gust of air knocked her in the face, and instantly her eyes flicked upward. She scanned the sky east to west, as if searching for the faint glow of the galaxies she'd studied over the past six decades. Now, on the catwalk of Kitt Peak National Observatory's 2.1-meter telescope, she walked slowly toward a second door, without looking down. Vera savored the silence and the starlight.

It was close to six in the evening on a cool November night in 2007, and Vera and her longtime collaborator, astronomer Deidre Hunter, were about to awaken the telescope and turn it skyward. They planned to point the huge glass eye at a spiral galaxy called NGC 801 and another called UGC 2885, one of the universe's largest swirls of stars.[1] After one last look at the stars, Vera again disappeared into the darkened telescope dome, crossed it, as she'd done dozens of times before, and walked back to the control room. She removed her hat and gloves, smoothed her cropped, white hair, straightened her owl-eyed glasses, and sat down. Faint rumbling began outside as a massive metal garage door slid open, and the telescope's mirror cover was systematically removed, giving the highly reflective glass eye sheltered inside its first sight of the sky that night.

Then there was silence. Vera and Deidre pushed buttons, flipped through their observing logs and notebooks, and continually checked the telescope. Finally, about thirty minutes later, Vera spoke: "We are getting more distance than I did in 1980."

In 1980, Vera and her colleagues had been trying to find stars at the extreme fringes of spiral galaxies and measure those stars' velocities.[2] Those velocities—the speeds at which the stars whip around the galaxy—provide clues about a galaxy's structure and composition. They reveal tantalizing clues about the shape of a galaxy and its heft. Vera had become fascinated with galaxies in graduate school, and about a decade later, in the 1960s, when she started studying the stars and hot gas of our nearest spiral neighbor, Andromeda, she spotted something odd. Nearly all of the galaxy's stars and gas traveled around the spiral's core at roughly the same speeds. The stars and hot gas sitting far from the spiral's center were moving much faster than she expected.[3]

Vera wasn't sure what to make of what she saw. The data seemed to defy the laws of physics. That's because astronomers had thought a galaxy's stars and gas would swing around the spiral just the way planets circle the sun: the closest objects move the fastest, while the ones farthest from the star's gravitational grip move more slowly. A spiral galaxy's stars, many astronomers assumed, should behave similarly, with the stars closer to a galaxy's core traveling much faster than the stars farther out. But that wasn't what Vera saw. Instead, Andromeda's outermost stars moved around the galaxy just as quickly as the stars closer to its core—very unusual behavior.

Vera wasn't the only person who saw that stars and gas in galaxies didn't behave the way astronomers had expected. Radio astronomers, who studied the heavens in radio wavelengths, were also observing faster-than-expected speeds for galaxies' gas.

Nature seemed to demand that astronomers rethink the content of the cosmos.

To explain the swift speeds of stars and gas that Vera and others had observed, the universe would need more matter. Astronomers didn't know what it was or really where it was, but the speedy stars and gas suggested that some form of additional matter did in fact exist. This extra matter was called "dark matter," a nod to astronomers in the 1920s and 1930s who first noticed that stars in galaxies and galaxies in clusters didn't move at the speeds Isaac Newton's law of gravity predicted; they moved much faster. To explain the fast speeds of stars in the Milky Way, Dutch astronomer Jan Oort in the 1920s had suggested that our galaxy might have dark matter tugging them along. A few years later, an astronomer at Caltech, Fritz Zwicky, also suggested that galaxies needed extra mass; without it, the Coma cluster—a clutch of more than a thousand galaxies sitting some 322 million light-years from Earth—should fly apart. Yet that wasn't what astronomers saw. Instead, the galaxies in the cluster stuck together. Around that same time, as World War II loomed, Horace Babcock, a graduate student at the University of California, Berkeley, also made observations that hinted at the existence of dark matter. He showed that Andromeda, which Vera would study decades later, might need some extra matter, again to explain the fast speeds of stars circling the galaxy's center. The universe, the studies seemed to suggest, wasn't what it seemed. The universe needed more matter. Those clues pointing to the need for additional matter to exist failed to capture astronomers' attention, and so the hints that dark matter existed went unheeded.

More evidence for dark matter emerged again in the 1950s and 1960s from radio astronomers. Then Vera came along and, drawing on Oort's work, calculated that the Milky Way's stars seemed to move too fast for Newton's law of gravity. When she and Kent Ford pointed a telescope at Andromeda, they saw the same thing. So did radio astronomer Mort Roberts, while Seth Shostak, a graduate student at Caltech, and his PhD adviser, David Rogstad, also spotted gas moving too fast in a few other galaxies. Shostak even took a crack at explaining the fast gas, positing in 1971 that the galaxies he studied had to contain some form of invisible matter. Collectively, the radio and Vera's optical observations captured theorists' attention, leading them to run computer simulations that showed that for galaxies to develop their spiral shape, they did have to have matter we could not see. Specifically, they had to have a halo, or cloak, of subluminal matter—matter so faint our telescopes at the time couldn't detect it. And there was more. The simulations said there had to be ten times more subluminal matter than observable matter for galaxies to take their spiral shape. The case for the existence of invisible matter was beginning to grow, yet some astronomers still had their doubts. They questioned the radio astronomers' data, and they questioned Vera's too.

And so Vera and Kent went back to the telescope, as did the radio astronomers. And again and again, she and the others found the same thing: the stars and gas in galaxies were moving too fast to follow Newton's law of gravity. Even still, astronomers had doubts. They were being asked to believe in something they could not see, an exceedingly hard sell for scientists. And, they were being asked to overturn an assumption about galaxies that they'd held for decades—again no easy feat.

Vera hated scientific assumptions. She looked for every opportunity to invalidate them. She did, however, have a few personal presumptions that she used to navigate her life and her research, ones that drove her to test some of the long-held conventions of the astronomy community, conventions propagated mainly by men. Vera's presumptions were that half the brains of the world belong to women; women could solve a scientific problem just as well as any man; and we all need permission to do science, women especially so. She had the brains, she probed questions about the cosmos few else would, she could solve exceedingly complex problems, and she'd earned astronomers' permission to do so, so she resolutely set to the task of testing astronomers' assumptions about how stars and gas move around galaxies. Vera decided she'd look at dozens of galaxies to see if the stars and gas behaved similarly to what astronomers expected or if the stars and gas in those galaxies defied assumptions.

In ten, then twenty, then forty, then sixty galaxies, she and Kent found stars that moved too fast to follow Newton's law of gravity. Her work defied astronomers' assumptions about galaxies and their stars and gas. Slowly the astronomy community began to acknowledge that there had to be more to the universe than meets the eye. By 1980, when Vera was at the telescope pushing to find the farthest young stars in UGC 2885, a consensus was beginning to emerge about her work: the data were all but undeniable. All you had to do was look at the data, she said, and you could see that the speeds of stars and gas stayed steady, even farther and farther from a galaxy's core.

The unexpected speeds demanded an explanation.

They could be explained, astronomers finally conceded, if galaxies had large amounts of matter we cannot see—dark matter. It was Vera's perseverance that finally helped to convince the scientific community of the existence of dark matter. Her work was pivotal to redefining the composition of our cosmos, revealing that most of it is some mysterious form of matter that, even today, we do not understand.

1 Stellar Tales

Eleven-year-old Vera Rubin—Vera Cooper, then—stared at an imaginary line running down the bed she shared with her sister, Ruth, then rolled over, defeated. She was the younger of the two and was told she couldn't sleep next to the small row of windows that lined the inner wall of the bedroom and fortuitously faced north. But even from the inside edge, the starlight caught Vera's attention; she was mesmerized. Every night, she'd crawl over Ruth to get a better view of the sky. "I was very angry at her," Ruth said, because she'd "wake me up."[1] But Vera couldn't help it. "There was just nothing as interesting in my life," she said, "as watching the stars."[2]

Vera and her family had recently moved into a rented three-bedroom townhouse in Northwest Washington, DC. At first, she and Ruth had separate rooms, but soon the family decided they wanted a den, so they converted Vera's room into that shared space. Vera moved in with Ruth, a move that ultimately spurred Vera's love of the stars, much to her sister's dismay, at least back then.

Vera was born Vera Florence Cooper on July 23, 1928, at Temple University Hospital in Philadelphia to Rose Applebaum Cooper and Philip Cooper. Philip, an emigrant from Poland, was born just before the turn of the twentieth century as Pesach Kobchefski in the town of Vilna, now the capital of Lithuania. Philip's father was in the leather business, and around 1905, he moved to Gloversville, New York, to establish himself in the United States before sending for his family a year later. Philip, who quickly became known as Pete, had an older sister and two younger brothers. All of them lived a comfortable life in upstate New York, where the family stayed until Pete was a junior in high school. Then, leaving him behind to finish his schooling, the family moved to West Philadelphia to open a store where they sold leather goods, such as ladies' gloves, and to set up a leather tannery. Pete joined his family in Philadelphia a year later, after finishing high school, and enrolled at the University of Pennsylvania, probably the first person in his family to go to college, Vera said. "He was awfully smart. He was very talented mathematically," and there was never any question about whether he would go to college; it was certain he would.

Vera's mother's family also emigrated to the United States, just before the turn of the twentieth century. Her grandmother left an apple orchard in Bessarabia, in what is now Moldova and part of Ukraine, to travel to Pennsylvania, according to the stories she told her grandchildren. She "enthralled [us] as children [with] tales of her crossing the ocean essentially in steerage and not eating because the food wouldn't be kosher," Vera said. Her grandmother nearly starved on her voyage to America and probably survived only because one of the ship's officers regularly brought her fruit.[3] "She arrived in Philadelphia . . . I think at about the age of sixteen and within a few years, had married someone who had also come from near where she had come from. He was a tailor. He hand-sewed and finished things for John Wanamaker," a famous American merchant, Vera explained. Her mother, Rose, was born in 1900, grew up weren't children's books or toys. "There were my father's college books, I remember, and my uncles were certainly very literate, and there were things around, but not really for children," Vera said.

Perhaps sensing this lack of engagement for his daughters, Pete built them a dollhouse. He had gotten the plans from *Good Housekeeping* magazine, and board by board, he constructed an elaborate colonial mansion, complete with working lights and a working radio. It was a delight for Vera and Ruth to play with—a deeply cherished family toy. Ruth and Vera also loved listening to college football games on the radio, and when they were not playing with the dollhouse or taking in the play-by-play of a game, Vera often went to the Franklin Institute, a science museum in downtown Philadelphia. "I did sort of fall in love with the Franklin Institute," she said. "They had these cones that you put sand in, and they were suspended on chains from the ceiling." As they moved, the cones made Lissajous figures, graphic curves that describe harmonic motion, that got smaller and smaller. "I could have spent a day in front of those," she said. The institute also had a "very large kaleidoscope that you actually walked into. It had mirrors on all sides and lights coming in. . . . And I got very interested in kaleidoscopes."

But just as Vera was starting to explore the delights of the Franklin Institute, the family moved: Pete had taken a job overseeing construction of a state facility in central Pennsylvania. After years of sharing a home with their parents, the Coopers relocated to Selinsgrove, a borough northwest of Philadelphia nestled along the Susquehanna River. They rented a home from a college professor who usually worked at Susquehanna University but was on sabbatical. There, at that house, in the attic, Vera found a set of *Book of Knowledge* encyclopedias. She had not had books like that at her grand-parents' house, and so she was delighted to delve into them, recalling that from them, she learned to press flowers in sand and possibly to build a kaleidoscope. "I made a kaleidoscope out of my mother's icing squeezer," she said. She screwed wax paper with little pieces of material onto the front of the aluminum icing cylinder, then cut three pieces of glass to make the optical part of the device. It was the start of her tinkering with lenses and light.

The Coopers stayed six months in Selinsgrove. Then Pete was offered a job at the Department of Agriculture in Washington, DC, and the family moved again when Vera was ten. It was 1938.

For the first few months in the city, she attended H. D. Cooke Elementary School, near the Adams Morgan neighborhood, where she and her classmates worked in groups studying different subjects rather than sitting in desks, just listening to a teacher. "We studied South America. We made an enormous relief map, and we electrified it and put little lights in places and strings up to the board behind it and identified things. I had never done anything like that in school," she said. "It was just phenomenal; I just loved every minute of it. And I think about that time I decided I was going to be a kindergarten teacher because I liked all the cutting and the pasting and the 'making.'"[4]

Not long after moving, the Coopers relocated north to a townhouse on Tuckerman Street, near Takoma Park, where Vera looked out the windows along the wall of her shared bedroom and became captivated by the stars. Eager to learn more about science and the night sky, she went to the local library, where she read about great science legends of the past: Isaac Newton, Benjamin Franklin, and Franklin's cousin Maria Mitchell, whose story fascinated Vera because of the thrilling astronomical discoveries she made and because she was a woman.[5]

Born in Nantucket in 1818, Mitchell was raised a Quaker, a religion that promoted, even then, that girls and boys deserve an equal opportunity to education. Mitchell attended several schools and was also tutored by her father, a public school teacher interested in astronomy and mathematics. At the time, women were encouraged to study science, and Mitchell's father, William, noticed that Maria had a knack for mathematics, so he had her help him with his surveying and navigation calculations. At age twelve, Maria helped her father observe a solar eclipse and use the data to calculate the location of the family's home. Vera was inspired: that was roughly the age she was when she started watching the stars. A few years later, Maria started to help sailors calculate coordinates for their whaling expeditions, and at age sixteen, she opened her own school in Nantucket, accepting both black and white students—a controversial decision in the 1830s, but one about which she was adamant.

Maria went on to become the first librarian at the Nantucket Atheneum, and it was while she was working there and helping her father in the evenings with his observations of the night sky that she strode, accidentally, into the professional astronomy world. On the night of October 1, 1847, Maria scanned the sky with a 3-inch telescope from the roof of the Pacific National Bank, where her father worked, and spotted a fuzzy blob that didn't appear in any of her astronomical logs. Observing the object closely and tracking its motion, she realized the spot probably wasn't a planet or a star but something more transitory, perhaps a comet. That thrill of Maria Mitchell's discovery—that instant in which she saw something no one else had seen—captivated Vera, making her wonder whether she'd discover something so awe-inspiring as a comet from her studies of the stars. As Vera read more on the discovery, she found Maria's excitement for the night sky encouraging, and so too was the lesson she shared on surrounding yourself with individuals who would champion your passion, not quash it.

That became clear when Vera learned what happened right after Maria spotted her unusual orb. Maria rushed to tell her father, and he told her that she should write a summary of her work and send it off to other astronomers to announce her discovery. But Maria was hesitant: she didn't think the scientific community would recognize her work. Her father disagreed. Determined that his daughter earn the recognition she deserved, William wrote to his friends and colleagues in the astronomy community. These were influential individuals such as William C. Bond, the director of the observatory at Harvard. At her father's colleagues' urging, Maria eventually wrote a short description of her discovery, submitting it in January 1848, under her father's name, to *Silliman's Journal*, a US-based publication. The next month, she submitted a second description, a calculation of the comet's orbit—a publication that would ensure she was given credit as the comet's original discoverer.[6]

Impressed by the work, Harvard's president at the time, Edward Everett, suggested that Maria apply for a gold medal from the Danish king for her discovery. In the early 1830s, Frederick VI, an amateur astronomer and king of Denmark, had announced that he would award a gold medal prize to the first astronomer to use a telescope to spot a comet too faint to be seen with the naked eye. Frederick VI died in 1839, after he had made the proclamation, but his successor and son, Christian VIII, chose to honor his father's wishes. On October 3,

1847, two days after Maria's discovery, Francesco de Vico independently observed the same comet as Maria and immediately reported it to European authorities. Since word of Maria's discovery hadn't yet reached Europe, the comet was named after de Vico, and he was awarded the Dane's gold medal for the comet's discovery. When word of Maria Mitchell's discovery finally reached the king, however, he had to choose whether to award her a medal. Frederick VI's original proclamation had said he would award the medal to the *first* discoverer of each newly identified comet, but Christian VIII had already awarded the medal for Maria's comet to de Vico.

King Christian ultimately chose to award Maria the medal, and he allowed de Vico to keep his, a decision that made Maria Mitchell famous, as the comet she discovered became known as Miss Mitchell's Comet. With her discovery, she joined the ranks of Caroline Herschel and Maria Margaretha Kirch, the only two other female astronomers to find comets. Maria Mitchell's work, however, did more than put her in the ranks of Herschel and Kirch. It also changed European astronomers' perception of American astronomy; they could no longer snub the discoveries made across the Atlantic.

Maria Mitchell's success didn't cease with her discovery of a comet. It, along with her subsequent observations, as well as her teaching prowess, earned her the first faculty appointment at Vassar College when the school opened in the 1860s. She was named professor of astronomy and also the director of the Vassar College Observatory, and she took her work there very seriously, hoping to invigorate the minds of the women she taught, no matter the future path they chose. "When we are chafed and fretted by small cares," she supposedly told her students, "a look at the stars will show us the littleness of our own interests."[7]

Maria's story stuck with Vera, especially the idea that watching the stars could become a career and that Vassar had an observatory and astronomy classes. The story, along with her nightly study of the stars, helped to shape her career aspirations. "As soon as I got interested in astronomy, I just decided that's what I was going to do for the rest of my life," she said. "I mean, how can you be here with these things [the stars and planets] around and not know about them?"

Vera's parents weren't oblivious to their daughter's fascination with the stars, and they encouraged it most of the time. The late-night observing wasn't so well received, and Vera's mother once called to her as she left, "Vera, don't spend the whole night with your head out the window!"

Vera didn't always listen. She would sit up late into the night memorizing the motion of the stars as they moved across the sky. In the morning, she would draw from memory where the stars had traveled, and then she'd talk about what she saw. Vera also tried to surround herself with more people who encouraged her passion for the stars. One woman who indulged Vera's rapture with the night sky was Goldie Back, one of her mother's friends. Vera said that Goldie was one of the most influential women in her life.

Goldie had wanted to be an engineer, though the University of Pennsylvania would not permit her to study the subject, so she earned a teaching degree instead and moved to Washington to work at the Bureau of Standards. There, she met and later married Michael Goldberg, and the two, Vera recalled, lived a very academic lifestyle. They traveled to conferences and lectures, and they had a convertible and would take Ruth and Vera out into Virginia, where they could clearly see the stars. On those nights, the Goldbergs pointed out constellations, but Vera reiterated that she wasn't as interested in the mythological outlines that the stars made as much as she was in tracing the stars' motions. That fascination with the stars, not the constellations, might have seemed trivial at the time, but it was significant in the grand scheme of Vera's work, paralleling the way Albert Einstein, at a similarly young age, had imagined thought experiments that would later lay the foundation for special relativity.

To aid her studies of the stars and their motions, Vera decided she'd build a telescope. One day, she traveled alone to downtown Washington, DC, where she "picked up this free cardboard tube that linoleum came in and brought it home on the bus." She bought a 2-inch lens from Edmund, a discount optical supply company, and, with her father's help, she put together her tool to better study the stars. Vera's father constantly indulged his daughter in these intellectual pursuits. In fact, anything she needed to quench her curiosity for the world, her father would help her get or do. "He had a very analytical way of looking at things, and I enjoyed that very much," she said. "I think that was a very large influence."[8]

Aligned with his relentless will to help his daughter learn, Vera's father took her to the local amateur astronomy club meetings. She had wanted to join on her own, but Pete wouldn't allow it.[9] He thought she was too young, so he joined and took her to the gatherings to listen to talks by notable astronomers such as the Harvard College Observatory Director Harlow Shapley. In the late 1910s, Shapley famously repositioned the location of the sun in our galaxy, moving it farther out from the Milky Way's center.[10] He also argued that there were no astronomical objects outside our galaxy, that the Milky Way was the entire universe.[11] The Mount Wilson astronomer Edwin Hubble proved Shapley wrong in the 1920s, and so Shapley began to map galaxies, studying how they clustered together and were distributed around our own galaxy. It was this work, which he might have mentioned in his talks, that drove others to study the stars in our galaxy and galaxies in clusters and sparked the thought that the universe may have more matter than we can see.

2 Dark Matter's Debut

As Vera peered out the window night after night watching stars swing across the sky, astronomers were slowly becoming aware of the existence of dark matter, though they didn't really know it then. The idea of some unseen substance was so strange, so foreign, that it would need someone like Vera—through the special way she formulated her research—to champion dark matter's existence.

So at first, the idea that it was out there in the universe was largely ignored. An inkling that it might exist cropped up in the 1920s in Dutch astronomer Jan Oort's study of the Milky Way, and then emerged again a few years later in a seemingly wild idea posited by Fritz Zwicky. A bit of a notoriously irascible genius at Caltech, Zwicky was studying a group of eight galaxies, which were part of the Coma cluster, in the early 1930s, when he noticed they didn't behave the way Newton's law of gravitation suggested they should: they were moving too fast. To explain the galaxies' swift velocities, he reasoned that they had to have large amounts of unseen mass, which would give them a gravitational grip that kept the distinct cities of stars neatly tucked together. Without this invisible matter's gravity, Zwicky surmised, the speedy galaxies would have been zipping out of the cluster altogether. But they weren't, an observation that led him to conclude that the cluster must contain what he called *dunkle Materie*—dark matter.

Zwicky's use of the term in 1933, when Vera was just five years old, has often been cited as the first mention of dark matter in a scientific paper. But it is not. The moniker had been introduced into the scientific literature years before, though it is fair to say now that Zwicky was onto something with his claim of a mysterious mass influencing the movements of the galaxies. Many astronomers, including Vera, would not recognize the significance of Zwicky's work for several decades. Why? Because it's hard to believe in something we can't physically see.

The debate about unseeable stuff in the universe is centuries old. In fact, humans were speculating about whether there was more to the universe than meets the eye possibly as early as the fifth century BCE. That's around the time the philosopher Philolaus hypothesized that a planet counter to our own, called Antichthon, orbited an unseen "central fire" opposite Earth. And Philolaus wasn't alone in his thinking. The atomists, who argued that all matter was made of atoms—infinitely numerous, indivisible building blocks that existed in endless space—also argued that the universe contained an infinite number of unseen worlds.

Aristotle disagreed. His model of the universe countered the atomists' view, placing Earth at the center of a cosmos that did not change, and so, he said, unobservable matter did not exist. His depiction of the Earth-centric universe dominated cosmology in the Western world for centuries, at least until Nicolaus Copernicus's notion that the sun, not the Earth, was at the center of the cosmos caught on. Around that time, Giordano Bruno also dared to argue that other worlds did exist, that the stars in the sky were distant suns orbited by their own planets, and that the universe was infinite, with no center. The existence of unseen celestial objects seemed to be a bit of a game of ping pong back then.

Galileo Galilei would change that. Pointing his telescope to the sky in the early 1600s, he was perhaps the first to show that celestial bodies imperceptible to the human eye did in fact exist. His work ultimately revealed that the universe was a physical construct, not merely a philosophical idea. Eyeing the night sky through his telescope, he revealed previously unseen moons orbiting Jupiter and an uncountable number of stars comprising the galaxy's milky appearance. Galileo's observations showed that "the Universe may contain matter that cannot be perceived by ordinary means . . . and that the introduction of new technology can reveal to us forms of matter that had previously been invisible," astrophysicist Gianfranco Bertone and cosmologist Dan Hooper wrote in a 2018 review of the history of dark matter.[1]

The telescope was not the only tool astronomers used to detect the existence of objects in the night sky. Another was Isaac Newton's laws of motion and his law of gravitation. In 1783, John Michell used Newton's laws to reason that if gravity applied to light, there could be massive, unseen objects in the universe with so much gravitational tug that even light could not escape them. It was the first notion of the objects we now call black holes. About a decade later, Pierre-Simon Laplace came to a similar conclusion, suggesting black holes existed.

It wasn't until 1844, however, that astronomers started to truly grasp that invisible astronomical objects could reveal themselves solely by their gravitational influence. This became apparent when German mathematician and astronomer Friedrich

Wilhelm Bessel noticed small oddities in the motions of the stars Sirius and Procyon. The only way to explain the deviations was "if we were to regard Procyon and Sirius as double stars," meaning they were each orbiting another object—an invisible one. If that were true, then "their change of motion would not surprise us," Bessel said. He referred to the object tugging on Sirius as its "dark companion."[2]

The discovery of Neptune is another example of detecting an unseen object by its gravitational grip and then finding it in the night sky. In 1846, the French astronomer Urbain Le Verrier and the English astronomer John Couch Adams separately reasoned that oddities in the motion of Uranus suggested that there was another unseen planet in the solar system. Le Verrier wrote a letter calculating the coordinates of this invisible planet's location on the sky based on Uranus's unruly motions and sent it to German astronomer John Galle. Upon receiving the letter, Galle, who was working at the Berlin Observatory, asked for permission to check Le Verrier's coordinates. Sure enough, a planet was spotted not far from where Le Verrier said it should be.[3]

Emboldened by the discovery, Le Verrier made a daring claim: a dark planet, he said, orbited the sun closer in than the planet Mercury. He invoked the gravity of this dark planetary companion to explain an unusual feature of Mercury's orbit around the sun. The closest point of Mercury's orbit does not stay fixed in the exact same location with each trip around the sun; rather, the closest point slowly shifts with each trip. All of the planets' closest points of orbit do. Newton's equations account for this movement, called precession, which is a consequence of the planets' gravitational pull on each other. And they worked for every planet—except Mercury.[4] A dark planet, Le Verrier reasoned in 1859, could account for the discrepancy.

Since this unseen planet was supposedly closest to the sun, Le Verrier called it Vulcan, after the Roman god of fire.[5] And because his prediction of Neptune had been correct, Le Verrier's contemporaries didn't doubt the dark planet's existence. They began to search for it, alas to no avail. That's because there was no such planet. Decades later, Albert Einstein's general relativity predicted Mercury's precession precisely with no need for any missing mass.

As they continued their studies based on gravity, astronomers also started using photography to study space, and as they let their camera shutters open longer and longer, their images offered a deeper view of the stars, nebulae and everything in between. "These exposures revealed objects that were too faint to be seen by the eye even with the largest telescopes," astronomers David Malin and Dennis Di Cicco wrote in 2009. "The dramatic transformation of photography from a recorder of the visible to a detector of the unseen, opened a window onto a universe that was much bigger and more mysterious than anyone had imagined."[6]

One of the mysteries that astrophotography began to reveal was the way stars were unevenly scattered across the sky, with dark spots sitting right in the middle of bright, dense clumps. Such dark spots, some astronomers argued, existed because there weren't any stars in those spaces. Other astronomers weren't convinced; they reasoned that the dark spots were "absorbing masses in space, which cut out the light from the nebulous or stellar region behind them."[7]

Spectroscopy, a measure of the intensity of light of an object emitted over a range of energies, would help astronomers solve the mystery of some of the dark spots in space. Each element of the periodic table can produce a unique series of bright lines of light, or they can absorb light at those same regions, and those lines reveal what a star is made of, its temperature, and its size. The lines also revealed that the dark spots in space were not cavities at all but were gaseous in nature, Father Angelo Secchi, director of the Roman College Observatory, wrote his memoir in 1877.[8] The discovery was yet another clue that what we see with our eyes doesn't always tell us the entire story.

Another part of that story of what we see in space is tied to those spectral lines. Those lines, or spectra, aren't only good for identifying what comprises a star or gas cloud. They can also reveal whether an astronomical object is moving closer to Earth or farther away from it, along astronomers' line of sight—motion that astronomers call radial velocity. Measuring stars' radial velocity would become more and more important as the astronomical community started to seriously discuss the existence of unseen material in space. And so the conversation was shifting from dark stars, planets, and nebulae to the existence of some other type of dark material.

One of the first scientists to make the leap from dark stars and planets to dark matter, Bertone and Hooper explain in their review, was physicist William Thomson (better known as Lord Kelvin). He was the first to question whether it was possible to calculate the amount of dark material in our galaxy.[9]

Kelvin was working with the assumption that "many of our supposed thousand million stars, perhaps a great majority of them, may be dark bodies," as he described in a lecture delivered at Johns Hopkins University in 1904. To make his calculation, Kelvin assumed that the unseen stars were as bright and big as the sun in our solar system. He then thought of those stars as particles of gas influenced by each other's gravity and used the conceptual framework to conclude that many of the stars in the Milky Way may be "extinct and dark." The majority of stars, nine-tenths, he said, may not be bright enough for astronomers to observe with the technology of the time.[10] So there could be dark matter but, more likely, he concluded, there were many, many stars we just couldn't see yet.

Kelvin's calculations intrigued French mathematician Henri Poincaré, who was taken with Kelvin's use of the "theory of gases" to describe the physical properties of our galaxy. In 1906, Poincaré suggested that the amount of *matiere obscure*, the dark matter, in the Milky Way (then considered the entire universe) was probably equal to or less than that of visible matter.[11] He later argued, with colleague Henri Vergne, that based on the calculations, there may be no "dark matter," only perhaps dark stars obscured from our view, or that if there is dark matter, it is not more abundant than visible matter.[12]

Not long after Poincaré and Vergne started discussing Kelvin's assessment of dark matter in the galaxy, Estonian astronomer Ernst Öpik also began a calculation of the amount of invisible matter in the Milky Way. At the time, in the early twentieth century, scientists pictured the Milky Way as an amoeba-like blob of stars with a bright center surrounded by a flat disk—what's called the galactic plane. Using the speeds of stars and their distances from the galactic plane, Öpik calculated that dark matter in the galaxy didn't exist at all; instead, its mass was most likely in stars that should eventually be detected with newer, more sophisticated telescopes.[13]

Jacobus Kapteyn, a Dutch astronomer, added to the discussion. He saw the galaxy as a flattened disk with stars scattered through it like blueberries in a pancake; the sun was a blueberry close to the core of the pancake, and all of the stars, including the sun, circled the galaxy's center. In essence, he argued, the galaxy was rotating.

As Kelvin and Öpik had done, Kapteyn related the motion of the stars around the disk to their velocity dispersions, statistical dispersions of the stars' velocities about the stars' mean velocity, which give clues to the mass of a galaxy. Kapteyn then attempted to calculate the density of the galactic system by dividing the total gravitational mass by how many stars there were, including faint ones. The technique provided a way to estimate the mass of dark matter in the universe (since the galaxy was the universe). "As matters stand at present," Kapteyn concluded, "it appears at once that this mass [dark matter] cannot be excessive."[14]

British astronomer James Jeans reanalyzed data on the vertical motions of stars near the plane of the Milky Way that same year, in 1922, and concluded that some dark matter probably did exist near the sun, perhaps two dark stars for each bright one.[15] A decade later, around the time Vera would have been four years old, a seemingly more definitive suggestion of the existence of dark matter in the Milky Way emerged.

The evidence came from Kapteyn's student Jan Oort. He too looked at stars' velocities—the speeds of stars moving in a specific direction—and found that there was not enough mass in the stars themselves to tug them all along at the speeds they were measured to be moving. The mass of the stars alone fell short of the mass indicated by the stars' velocities by about 30 to 50 percent, he wrote.[16] Oort and many others still thought that such mass was probably made of faint stars astronomers couldn't yet detect with telescopes, along with what he called "nebulous or meteoric matter."

Dark matter, Oort said, was "probably less than the visible stars, possibly much less."[17]

Again, the idea of dark stuff in the galaxy, and the universe, felt like a game of ping pong. Back and forth the arguments went on whether it existed and just how much was really there. Then Zwicky came along and looked at the Coma cluster. When he found it needed more mass than can be seen, it was no surprise that he used the German phrase *dunkle Materie* since that is what had appeared previously in the scientific literature. What was unique about Zwicky's work, however, was that he was looking at the radial velocities of galaxies, not the velocities of stars in a single galaxy. By this time, Edwin Hubble had shown that galaxies other than the Milky Way exist. Through a correlation in the relationship of the velocities of galaxies and their distances from Earth, Hubble and his assistant Milton Humason also showed the universe is expanding. Knowing this, Zwicky collected the radial velocities of various galaxies published by Hubble and Humason in 1931, and noticed large variations—of more than 2,000 kilometers per second—in galaxy velocities in the Coma cluster. Hubble and Humason had noticed such large velocity dispersions too, but they didn't pursue the matter. Zwicky did: he applied a mathematical equation that relates the total kinetic energy of a system, in this case the Coma cluster, to its total gravitational potential energy. The equation is part of the virial theorem, which says that the potential energy of a system must equal the kinetic energy, within a factor of two.

Zwicky calculated the potential energy of the cluster using a rough estimate of its size, a million light-years. Then, he calculated the rough kinetic energy of the system using an estimate of the number of galaxies the cluster contained, roughly 800. And he assumed an average mass of each galaxy of a billion times the mass of the sun, as Hubble suggested. When he made his calculation on the cluster, he concluded the velocity dispersion of the galaxies should be about 80 kilometers per second. That was way off what Hubble and Humason reported in their data, leaving Zwicky to deduce that if the galaxies in the Coma cluster were to remain together, the cluster would need ten to one hundred times more mass than what he'd calculated from the luminous matter of the galaxies alone. "If this would be confirmed, we would get the surprising result that dark matter is present in much greater amounts than luminous matter," he wrote in a Swiss physics journal in 1933.[18]

Intrigued with Zwicky's work, Sinclair Smith, a Caltech graduate student, took up the question of the mass of galaxy clusters. He'd built a spectrograph and used it to study the velocities of galaxies in the Virgo cluster. Based on the velocities, and assuming the galaxies orbited the center of the cluster, Smith calculated that all of the galaxies combined had a mass of 100 trillion suns. He then went on to divide 100 trillion by 500—the number of galaxies in the cluster—and estimated that the average mass of each galaxy in the cluster was roughly 200 billion suns, much more than the billion suns Hubble had estimated. The extra mass, Smith suggested in 1936, could be material floating uniformly throughout the cluster or in "great clouds" around each galaxy.[19] Either way, the cluster needed more mass than what could be seen to keep it together.

In the same year that Smith reported his results, Hubble published *The Realm of the Nebulae*, a book of lectures that he'd given at Yale a year earlier. Hubble recounted his discoveries that there were other galaxies outside the Milky Way and that based on the velocities of those external galaxies, the universe was expanding. He also discussed the notion of dark matter, which arose because of the measurements showing that stars in galaxies moved too fast if only taking into account observable matter. The discrepancies in the stars' speeds versus the matter we could see, Hubble wrote, was a "real and important" problem, one that might resolve itself, since the masses calculated were limits—upper limits of dark matter in galaxies and lower limits of dark matter in galaxy clusters. Still, the situation was confusing: Did dark matter exist or didn't it? Astronomers couldn't say for sure.[20]

Zwicky, meanwhile, was redoing his work on the Coma cluster, trying to find a better way to calculate the mass of each galaxy. He still turned to the theorem he'd used in 1933 but tried different numbers for his calculations. This time he estimated the Coma cluster had 1,000 galaxies and a radius of 2 million light-years. With a velocity dispersion of 700 kilometers per second, his mass for the entire cluster was 45 trillion suns. Each galaxy, then, weighed in at roughly 45 billion suns, still a bit heftier than Hubble suggested and a bit more than what astronomers expected based on calculations from the starlight alone in the cluster. Adding up the masses of the stars from the light that could be observed, astronomers thought the Coma cluster should tip the scales at roughly the mass of 85 million suns. Zwicky's estimate was 500 times greater.

This was a clue that dark matter might make up much more of the cluster, and possibly the universe, than normal matter. Of course, as Zwicky noted in 1937, in order to derive the mass of galaxies from their luminosity, "we must know how much dark matter is incorporated in nebulae in the form of cool and cold stars, macroscopic and microscopic solid bodies, and gases," a nod to the idea that dark matter might be cool bodies that astronomers couldn't detect.[21]

Zwicky was on to something. And a year later, more evidence emerged to suggest he might be right—that astronomers didn't understand everything going on in the universe. This time, though, the oddity came from individual galaxies. It was 1938, and University of California, Berkeley graduate student Horace Babcock was using Lick Observatory's 36-inch Crossley reflecting telescope to observe the center of the Andromeda galaxy. He recorded the spectra, the wavelengths of light emitted, from a few stars close to the galaxy's core. Taking those spectra was tedious; Babcock had to observe each star anywhere from seven to twenty-two hours, but he diligently made the measurements. With the data, he could compare the velocities of the galaxy's stars to their distance from the galaxy's center and sketch out the galaxy's rotation curve. The curve didn't seem to drop off.[22]

This was work that would set the foundation for Vera to lead a revolution in astronomical thought, but at the time she was just nearing the age when she began watching the stars out her bedroom window. Little did she know that along with the ways stars moved, she too would become fascinated with what she couldn't see in the cosmos. Still, the idea of invisible matter in the universe was tough to grasp. Babcock's data and Smith's and Zwicky's calculations were not enough to convince the scientific community to take seriously the existence of dark matter. In fact, many researchers doubted the implications of the galaxy cluster work all together. Swedish astronomer and cosmologist Erik Holmberg even wrote in 1940 that "it does not seem to be possible to accept the high velocities [in the Virgo and Coma clusters] as belonging to permanent cluster members, unless we suppose that a great amount of mass— the greater part of the total mass of the cluster—is contributed by dark material distributed among the cluster members—an unlikely assumption."[23]

Astronomers had hit pause on the dark matter idea, and it stayed that way until well after World War II.

3 Waiting on the Stars

As World War II began to consume Europe, China, and eventually the United States, Vera stayed focused on the stars—at least when she could. She worked on weekends, school holidays, and summers in the Selective Service office in DC filing papers. "It was a job I detested. . . . They were the most horrible hours of my life. I would watch the clock. It was not interesting," she recalled. "My mother, who knew how much I detested it, said it was good for me to see what it was like to do something I didn't like. It was close to unbearable."[1] Though Vera hated working in the office, she escaped the monotony of the job through reading. She devoured James Jeans's *The Universe Around Us*, as well as Arthur Eddington's *The Expanding Universe* and *The Internal Constitution of the Stars*. Jeans's book, she said, had "wonderful ideas . . . ideas about whether, when you look in one direction, you can see light from stars in other directions. It was those kinds of concepts that really fascinated me more than the everyday, conventional astronomy."[2] Her flair for asking unconventional questions in science flourished early in her life.

By the time she was in high school, Vera knew she wanted to be an astronomer. "I didn't know a single astronomer [personally]," she said, "but I just knew that was what I wanted to do." She of course had heard of Maria Mitchell and, with her father, attended talks by Harlow Shapley and others during DC's amateur astronomy club meetings, but she knew no professional astronomer well enough to ask him or her to nurture her love of stars. Most of her high school teachers weren't much help either.

Vera attended Coolidge Senior High School in Northwest Washington, DC, and there she had a phenomenal math teacher named Mr. Lee Gilbert. "The first geometry test he gave us, he asked us to prove something that we didn't yet know enough to prove. And he sort of walked up and down the aisles, watching what you did," she recalled. "He made us think; he really made us think. On our feet. He would call you to the blackboard, and he would say we had to talk as if people were blind and draw as if they couldn't hear or something. . . . That is probably the best thinking I ever did in school." Vera also took Gilbert's algebra class. "He would say, 'Always put off to tomorrow what you don't have to do today because tomorrow you may not have to do it.' That had to do with multiplying factors," she said. "He was a remarkable man. He really was the best teacher I met in my entire career."[3]

Gilbert seemed to be the only one to inspire Vera and encourage her. Others at the school weren't so remarkable, which Vera admitted was partly her fault. "My sister, who had been two, two and a half years ahead of me, was very smart and had done very well, and all the science teachers had liked her," Vera said. "I was her sister, and she had done so well. And I really was terribly independent, and that just made me want to have nothing to do with these teachers or their expectations."

There were other barriers to success aside from Vera's sibling rivalry. "The physics class was a big macho boys' club" in which the teacher, Mr. Himes, and the other boys ignored the few girls who were there. "I think it was one of the worst misfits of my life. The physics labs were a nightmare," Vera said. It was a firsthand experience of the way in which the school reinforced gender norms of the time, encouraging men in the sciences and math and women in languages and the arts. There was even a boys' wing housing the boys' gymnasium, the wood shop, and the metal shop. Vera wanted to take mechanical drawing, a class that was given in the boys' wing. But the prospect of attending alone terrified her, so she persuaded a girlfriend to join her. Vera ended up loving the class.

Even with such small victories, Vera's path to becoming an astronomer would continue to be challenging. One of her biggest decisions at the time was where to attend college. Her sister went to George Washington University, and all of her cousins had gone to college too, usually close to home. "I had not known anybody that went away to college. I really wanted to study astronomy, and I didn't know of any place I could do it nearby," she said. She also said she dreaded the thought of studying physics after her experiences with Mr. Himes, so she was now looking for a school somewhat nearby where she could focus on astronomy. To help her decide where to apply, Vera turned to the library, reading up on schools with astronomy programs. She also recalled her parents' friends, Goldie and Mike Goldberg, had recently been to a mathematics society meeting at Vassar, which, Vera said, "impressed me no end, I mean, the fact that they would go to that college." Coupling that impression with her fond memories of reading about Maria Mitchell, Vassar was one of Vera's top college choices. She also applied to Swarthmore and the University of Pennsylvania. She considered Radcliffe and came close to applying, she said, but decided not to since she thought her chances of getting a scholarship would be slim.

Vera was accepted at the University of Pennsylvania and Vassar. Swarthmore turned her down—a story that has become family lore. She recalled meeting with one of the deans of the school during an interview in DC. "The interview was a total disaster, absolute disaster. And I knew it at the time," Vera recalled. It was a disaster, she said, because the woman never took her seriously, especially not her desire to become an astronomer. When Vera told the woman she was interested in astronomy and painting, as a hobby, a relic of her cutting and pasting love during childhood, the woman asked Vera if she'd ever thought about a career painting pictures of astronomical scenes. "That line became a joke in our family," Vera says. "If you want to say something funny you say, 'Have you ever considered a career in which you paint astronomical scenes?'"[4]

Vera was not surprised she wasn't accepted at Swarthmore. She was, however, surprised and thrilled to get a scholarship to Vassar, from Washington, DC, alumnae, whose funds sponsored a student from the District to attend. She was ecstatic at heading to a college known for its astronomy program for women. Of course, not everyone celebrated the success as Vera did. Her school counselor marveled at her acceptance and her winning a scholarship and muttered, something like, "with your French scores." Vera ignored her. She was more interested in telling her physics teacher, Mr. Himes. When she finally said she'd earned a scholarship to Vassar, he quickly retorted, "As long as you stay away from science, you should do all right."[5] He, of course, didn't know that Vera was seriously interested in pursuing a degree in astronomy and had chosen Vassar because of its rich history of supporting women astronomers. Despite his negativity, Mr. Himes's rebuff didn't deter Vera, and it didn't even make her think to consider she'd be entering a field dominated by men. "I didn't think that all astronomers were male because I didn't know," she said. "I didn't even know how you became an astronomer." All she knew was that she was determined to become one—no matter what anyone said.

That determination was most likely seeded in the long, lackluster summer hours between her high school years that she spent filing papers at the Selective Service office. She spent the summer of 1945, before heading off to Vassar, in that office, and hour after hour it became increasingly clear to her that she did not want to spend the rest of her life sifting paper as work, especially after so much devastation as a result of World War II. "I remember, I would take a bus with my father to go downtown during the summers. And I remember the day the newspaper had the story of the dropping of the atom bomb," Vera recalled. "That's probably the most dramatic recollection I have of the whole war period. I was seventeen."[6]

That was early August 1945. Only a few weeks later, Vera was headed to New York's Hudson Valley, and when she arrived, she learned she was the only person in that year's freshman class at Vassar who had chosen to study astronomy. Again, it didn't strike her as odd, and because she had declared her interest in astronomy early, Vera was almost instantly connected with Maud Makemson, director of the Vassar Observatory.

Unlike Vera, Makemson had come late to astronomy. After teaching and working as a journalist, Makemson realized she had an interest in the night sky and taught elementary school while taking correspondence and summer courses to be eligible to attend the University of California. In 1925, she earned her bachelor's degree from UCLA and then went on to pursue a PhD, in which she calculated the orbits of asteroids, at UC

Berkeley.[7] She defended her dissertation in 1930 and joined the Vassar faculty two years later. When Vera met her, Makemson had recently finished books on Mayan and Polynesian astronomy and had been director of the college's observatory for nearly ten years. Makemson was the first astronomer who could mentor Vera.

Vera had a full load of classes first semester of her freshman year: astronomy, math, French, and English. "It was a wonderful environment," she said. Makemson taught Vera's freshman astronomy class, along with a history of astronomy course that met once a week. In the regular class, Makemson taught from William Skilling and Robert Richardson's *Astronomy*, first published in 1939, and had the students use 3-inch telescopes to make observations and drawings of the Orion Nebula and Saturn. "It was a fairly technical course," Vera said. But, she admitted, she wasn't too diligent with her lab work. In their notebooks, the students were supposed to regularly draw the horizon and note where the sun set. Vera was careful to track the first sunset and the last but did not watch or draw the sun set every night, she said. "Before [the assignment] was due, I would put a few Suns in because I would know how they should have crept along the horizon."[8]

While astronomy seemed to be a breeze, French wasn't so easy. Vera had had four years of the language in high school, and she was placed in the most advanced class at Vassar. She was in the class with young women who had visited and lived in France and spoke the language fluently. "That was probably the hardest course I've ever had in my life," she recalled. "I should have flunked it." She spent all her studying hours working on her French and recalled that for the final exam, she had to write an essay in French on King Louis XIV. He was called the Sun King, she wrote, but did not add much else to her essay. The professor said to her, "Is this all you know about Louis the whatever?"

Yes, it is. It really is, Vera thought to herself.

Despite Vera's limited ability to communicate in French her knowledge of the long-reigning European monarch, she didn't fail the class. "I don't think I ever got worse than a C, but it was terrible," she said. That challenging course, however, didn't change Vera's infatuation with her first year at Vassar. She was never lonely, never homesick, never tempted to catch a train home, she said. There was just too much to learn.

She did return home for Easter to visit her parents, see their new apartment, and secure a summer job at the Naval Observatory. While Vera was at Vassar, her parents had moved from their home in Northwest Washington to an apartment complex, Trenton Terrace, in Southwest DC. They had been renting the townhouse near Takoma Park during the war, and once the war was over, the owners sold it, so Vera's parents had to find somewhere new to live. It just so happened that Trenton Terrace was being built by a family Vera's parents knew, the Gerber family, "a local, very liberal, rich family," she said. "One of their daughters or nieces was a close friend of mine. And that's how my parents got an apartment there." It was a lovely flat, she said, though it didn't have northern-facing windows. But that was okay because Vera didn't spend much time there during the academic year. She was coming home for the summer, however, so she also went over to the Naval Observatory to check on a job she'd been offered there. "All the papers were signed. I was shown where I would be sitting. I was shown what I would be doing. And then somehow, sometime between then and the summer, I got a letter telling me there was no job," Vera recalled. "I really needed the money. And we were not too far from the Naval Research Lab. That's where this apartment development was."

Vera's father called around and helped his daughter secure a job at the lab, where her sister would be working in the library. When Vera finished her first year at Vassar and returned for the summer, she headed to NRL each day to help with a psychology experiment studying reaction times. The experiment was set up so that a participant watched a slot with a piece of paper that had a vertical line on it and put a pencil on the line. The line would move: it would jump from spot to spot, and the participant was supposed move her pencil so it was on the line each time it moved. Vera's job was to measure how fast participants moved their pencils and how far they overshot the line when it jumped. Not only did Vera record the data, but she also participated as a subject, making the experience "incredible great fun," she said.

Before she knew it, the summer was over, and in fall 1946, she headed back to Vassar. When she arrived, Maud Makemson was away on sabbatical. It was at that point that Vera said she began to experiment with the telescopes and other observing tools at the school. There were at least two telescopes, Maria Mitchell's original 12 3/8-inch telescope and a 5-inch telescope that had an instrument called a spectroheliograph that Vera could use to take images of the sun in a single wavelength of light. "I don't remember much about the spectroheliograph at all except it came into the side of the building. It was old. Everything seemed to be falling apart, and I could not get it going," she said. "But the 5-inch telescope was usable, and I talked one of my friends into helping me. So we would go out in the evening and take photographs, take plates, develop them. I mean [we] just sort of really played with this 5-inch telescope." She was "pretending" to be an astronomer, she said. She wasn't yet a real astronomer, in her mind, but dabbling in the field was something that she said she felt a bit freer to do without Makemson there. "[Maud] was a slightly formidable person, and I always felt slightly incompetent around her, truly. And so with her not there . . . I just sort of felt like I could play," Vera said. "The darkroom was not really in good shape, but I found chemicals, and I found plates, and I just sort of made myself at home."

Admittedly, Vera said she probably would have accomplished much more had Makemson been there to guide her, but it was a tricky situation, as being an astronomer at Vassar was not especially encouraged, even though Maria Mitchell had once led the program there. "I don't think I got enormous support for it. The feeling was that there were very few observatories and very few astronomers needed," Vera said. "I wasn't discouraged at all, but I can't say that I was overwhelmingly encouraged." In general, the women at the school were encouraged to study science, and there was an emphasis on physics and mathematics. "But astronomy, even there, was a very tiny department," she explained. "It was not a major department on the scientific scene." On top of the lack of encouragement, Vera wasn't yet interested in doing astronomy in a formal way. "I didn't want to set up a course where every Monday night I had to go photograph stars. I mean, that would have been harder to do for me," she said. "And I think had I asked for something from [Makemson] that would have been the kind of response." With Makemson gone, Vera was free to explore her own curiosities and teach herself to observe the stars.

Vera returned home after her second year of college, in the summer of 1947, and again had secured a job at the Naval Research Laboratory, this time working with astronomer Richard

Tousey. Tousey had recently measured the first ultraviolet spectrum of the sun using data from a spectrograph he built, which flew on captured V-2 rockets. Vera's job was to help calibrate and reanalyze the spectral data. She was given a transparent spherical bead that had served as the entrance aperture for the spectrograph—where light went into the instruments—and used it on an optical bench—a table of mirrors, light sources, and screens—so she could pass light through the bead and measure any distortions.[9] She also made a Fresnel lens, which was 18 inches in diameter, to concentrate light into a super-powerful beam. "Much of the summer I did optics," Vera said. Drawing on her course work at Vassar, she wrote a report on the solar spectrum, which, with her research at the lab, earned her two college credits. "That was the summer of 1947," she said, the summer she met Robert Rubin.[10]

During the war, Robert's parents had lived in Baltimore and commuted to DC for their government jobs. After the armistice, they moved to Trenton Terrace because they were vaguely related to the people who built it. It so happened that Vera's mother, Rose, and Robert's mother, Bess, met at the complex one day, talked about their children, and decided Vera and Robert should meet. It was around the Fourth of July weekend, and Vera had been away celebrating the holiday with the Goldbergs and their two children. When she returned home, she learned about Robert and his family and about the scheduled dinner. Legend has it, though, that the men, the fathers, in the family had already met, and not in a friendly way.

Pete, Vera's father, took the bus to work each day and saved a seat for a friend who was disabled and got on at a later stop. One day, Ben, Robert's father, got on the bus, and the only available seat was the one that Pete was saving. "Ben wanted the seat. Pete refused to give it up, and some sort of row ensued. That was the last they saw of each other until the dinner." That's when Robert, or Bob as he was called, Bess, and Ben showed up at Rose and Pete's apartment so that Vera and Bob could meet.[11]

Any hard feelings between the men apparently didn't prevent Vera and Bob from starting to date. In fact, Vera found herself quickly smitten with Bob. Earlier that year, she had heard a lecture at Vassar by the famed Cornell physicist Richard Feynman. Despite being an astronomy major and a part of the school's physics student group, she was too shy to talk to him, but, she said, "For me, Feynman was already a very romantic figure." At their first dinner, Vera decided to ask Bob, who was studying at Cornell, if he knew Feynman. Not only did Bob know him, he told Vera, he'd studied under him. "That immediately made Bob a very attractive figure," Vera said.[12] From that day on, they were together. They'd go to movies or baseball games, which Vera loved, and Bob hated, but still he went, because he loved being with her. At one of the games, Bob had chosen to wear a seersucker jacket. Sadly, the man sitting behind Bob and Vera was smoking and spilled ashes on the jacket, burning a hole in it. Only later did Vera learn the jacket wasn't actually Bob's; he'd borrowed it from a cousin. "We survived," she said, but there was a bit of embarrassment for Bob.[13]

As their feelings for each other developed, Bob started meeting Vera after work. "He had a summer job at the University of Maryland, and he took public transportation from Southwest Washington to the University of Maryland in College Park, about 12 miles," she said. "He started meeting me where I got off the bus, probably at some great personal sacrifice. It was hard to get there," she said. This continued throughout the summer, and then Bob and Vera went their separate ways— Bob to Cornell and Vera to Vassar.

Although they were physically apart, they'd talk about their work, about the latest breakthroughs in physics, and also Vera's life at Vassar, which, she said, wasn't as delightful as it had been before. When she returned in the fall of 1947, Makemson was back, "and things were incredibly difficult, almost to the point of tears," Vera explained. She was the only astronomy major in her class, and this was her final year. (She had chosen to do a program that would allow her to graduate in three years.) That year, Vera was taking an orbit theory course with Makemson and a few other women, which "was kind of fun," but large parts of it were heavily mathematical and not so much fun. She was the only student, however, in her celestial mechanics class, and "it just didn't work," she said, explaining that if she hadn't done the homework, she'd have to go to the blackboard and derive equations. "I would get to the blackboard, and I wouldn't do it right," she said. "It was terrible."

Though that class was unpleasant, Vera always looked forward to the weekends when Bob visited. "I think I must have decided pretty early, just like the astronomy . . . that this was someone I'd be happy to spend the rest of my life with," she said. Continuing to date wasn't always easy, though. "Neither of us had a car; we didn't live that way," so Bob would "devise schemes" to get from Cornell to Vassar. It "is not terribly far but also is not easy," so he would find someone that was driving there or partway there. He'd hitchhike to go see Vera.

By Thanksgiving, only a few months after they'd first met, Vera and Bob were engaged. With Bob in her life, everything seemed different. Her schoolwork seemed less important, even though she'd been serious about astronomy before, and her disinterest showed. "Whenever I wouldn't be able to do something, [Makemson] would say I hadn't been working hard

enough, and I should work harder. . . . It was very, very difficult," Vera said. "I mean, after being very, very serious, I think
she thought I was going to go off and get married and that was
the end of the astronomy."[14] Obviously it was not. But neither
Makemson nor Vera knew that then.

4 Threads of Research: A Rotating Universe and Radio Astronomy

Despite Maud Makemson's concern—one that was typical at the time—that marriage would be the end of astronomy for Vera, she chose to marry Bob. "It still was an era when people expected to get married as soon as they left college. That's what you did," Vera recalled. "I don't know if you were supposed to be actively looking for a husband, but it was sort of anticipated that is the way things would go."[1] Her decision to marry Bob helped her make her next important life choice too: where to pursue a graduate degree in astronomy. During her final year at Vassar, Vera had reached out to Princeton to inquire about its astronomy program but got a hasty reply informing her that the school didn't admit women, so she looked elsewhere. She applied to Harvard and was accepted, and she also considered following Bob to Cornell, where he was working on his PhD in physical chemistry. Like everything else, they discussed the options at length and ultimately Vera decided she would join Bob at Cornell. "We wrote our parents a letter, both sets of parents, telling them we wanted to get married . . . that we had talked about Harvard, but Bob only had two or three more years to go for his PhD, and it looked like I could have a comfortable spot," Vera said. "We just decided to go to Cornell."[2]

After making the decision, Vera wrote to Donald Menzel, the director of Harvard Observatory—and a speaker she'd heard during her days attending DC's amateur astronomy club—to let him know she wouldn't be attending Harvard. Vera recalled that he replied kindly, scribbling across a formal letter thanking her for letting him know—something to the effect of: "That's the trouble with you women. Every time I get a good one ready, she goes off and gets married."[3]

In May 1948, Vera graduated from Vassar, and she was scheduled to marry Bob on August 22 and then begin her graduate degree at Cornell in the fall. Clearly it was not the end of astronomy for Vera, as Makemson had assumed. In some ways, really, it was just the beginning. Vera of course didn't know it then, but as she left Vassar and prepared for her first day at Cornell, many of the different threads of scientific research that would influence her later work were starting to weave themselves together: the work on individual galaxies and groups of galaxies, and new ideas about the origins of the universe.

Vera even attended a talk about this new view of the early universe at the Cosmos Club in Washington, DC. She'd been home from Vassar on summer vacation, and the Goldbergs, friends of Vera's parents who had been influential in turning Vera into an astronomer, took her to the lecture. There, Ralph Alpher, then a graduate student of the cosmologist George Gamow, described how the chemical elements formed in the early universe, a process essential to shaping the composition and structure of the cosmos as we see it today. The idea proposed by Alpher, Gamow, and Robert Herman was that the universe got its start in what would eventually become called the big bang. The rough idea had been suggested in the 1920s and early 1930s by Georges Lemaître, who postulated that the universe had formed in the explosion of a primordial "cosmic egg."[4] Gamow expanded on that tenet, applying a more quantitative analysis, and with Alpher and Herman, described the primordial universe as a hot, dense medium filled with neutrons.

As the universe started to expand, the team suggested, the pressure of the gas dropped, and positively charged subatomic particles called protons began to form. When the protons paired with neutrally charged neutrons, the subatomic particle couple would be quickly ripped apart by high-energy packets of light called photons, which were abundant in the early universe. The universe was cooling, though, allowing protons and neutrons to finally start to stick together as the light packets lost energy and were no longer energetic enough to rip the subatomic particle pairs apart. Protons and electrons hooked up to make hydrogen, and then hydrogen captured a neutron to form deuterium. Then deuterium took on another neutron and an electron joined in too, generating helium, and on and on, as nuclear fusion continued, allowing heavier and heavier elements to form, the scientists said.[5] Doubts arose about how this fusion could create all the elements in the exact proportions we see today, but Alpher argued that the accumulation of subatomic particles was not always long-lived, as a type of radioactive decay, called beta decay, could shuffle the abundance of the elements around, at least until they matched with astronomers' latest estimates.[6]

Gamow, ever the prankster, loved this idea of the formation of the universe and the elements within it, and he also had a fixation on Greek letters (he called his physicist wife, Lyubov Vokhmintseva, Rho). The two loves were intertwined so much that Gamow decided that when the team submitted its paper on the origins of the universe, he'd add the name of physicist Hans Bethe, so the author list would be a play on the Greek letters alpha, beta, gamma. Herman, purportedly refused to change his name to Delter to represent the Greek letter delta. And Alpher, who did the heavy lifting on the research, was less than thrilled by the trick, but looked the other way, giving Gamow the go-ahead, which Alpher might have later regretted.[7]

All silliness aside, however, Gamow was serious about the results, even discussing them with Albert Einstein. Having long been a friend of Einstein, Gamow often took the train from Washington, DC, to Princeton to visit him. There, the two would go for walks, discussing their theories of the universe. On one outing, Gamow told Einstein about his theory of the origin of universe—that it was created almost instantaneously, out of nearly nothing, in an explosion. Einstein stopped, awestruck, in the street. All the cars stopped too, forming a line behind him—at least that's one of the legends they tell in Princeton of Gamow and Einstein's walks.[8]

Einstein wasn't the only one enthralled by the idea of the origin of the universe (which turned out to be mostly right, aside from the fact that the big bang led to the formation of *all* the elements). Hearing Alpher's talk, Vera was captivated too. She and Bob often discussed the idea, and it was probably in those lengthy conversations that Bob shared with Vera one of Gamow's other wild cosmic hypotheses: that the entire universe spun around some central point. "One of the most mysterious results of the astronomical studies of the universe lies in the fact that all successive degrees of accumulation of matter, such as planets, stars and galaxies, are found in the state of more or less rapid axial rotation," Gamow wrote in *Nature* in 1946. If galaxies, according to current theories, were "the result of gravitational instability of the originally uniform distribution of matter in space," he wrote, "we will find it very difficult to understand why such condensations are in most cases found in the state of rather fast rotation."[9] Based on the principles of physics and the current understanding of the universe, a total rotation of the entire system should be nonexistent, Gamow conceded. But, he reminded his readers, stars' spins probably originated from the rotating of the masses of gas in which they formed, and planets' spins originated from the rotation of the stars around which they formed. So then, he wondered, was it "not possible to assume that all matter in the visible universe is in a state of general rotation around some centre located far beyond the reach of our telescopes"?[10]

Gamow admitted the question was a "fantastic" one but argued that astronomers using the world's best telescopes at the time could collect the data necessary to answer it. That notion, that the question could be answered, would stick with Vera, and she'd begin to delve into it just as the first significant results from radio astronomy, another scientific thread essential to her future work, would begin to emerge. An early advocate of radio astronomy—the observation of the heavens in radio wavelengths—was Dutch astronomer Jan Oort, who had suggested dark matter existed in the Milky Way roughly two decades earlier. As Oort rose in the ranks at Leiden Observatory in the 1920s and 1930s, he grew adamant about turning the Netherlands into a leader in astronomy, and after World War II, he became particularly fixated on radio astronomy.

Influenced by his work with Jacobus Kapteyn, an astronomer who played a pivotal role, especially for fellow Dutch astronomers, in mapping the structure of the Milky Way, Oort continued to sketch out the structure and dynamics of the galaxy and had, back in the 1930s, begun to wonder if our galaxy was spiral-shaped, like a pinwheel. Other galaxies—which by then Edwin Hubble had shown were outside the Milky Way— were spirals, so it was possible, Oort thought, that the Milky Way was one too. His investigations of the galaxy would lay the foundation for Vera to later probe the dynamics of our swirl of stars and many others.

Our galaxy, however, was, and still is, hard to study because we're inside it, not looking at it from the outside, as we see other galaxies. Oort knew this, and so when he read a paper[11] by the pioneer amateur radio astronomer Grote Reber in 1940 about cosmic static and radio frequency emissions coming from the center of our galaxy, he began to wonder whether radio waves could offer astronomers an even crisper picture of our galaxy than what we could see in visible wavelengths of light.

Earlier research by Karl Jansky had shown radio signals coming from the Milky Way, but it was Reber's work that "made it quite clear [radio waves] would be a very important tool for investigating the galaxy, just because it could investigate the whole disk of the galactic system unimpeded by absorption," where a material between a celestial source of light and the observer absorbs some of the source's light, messing with its spectrum.[12]

The potential to see the galaxy in a new light was immensely intriguing to Oort, but he had to wait to act on the idea for many years. When he read Reber's paper, the Nazis were storming Europe, which led to the isolation of the Netherlands from the outside world; Oort's correspondence with his colleagues in the United States and Europe was cut off, and as Nazi power grew, the Jewish professors at Leiden were dismissed, and not too long after, Oort left his post in solidarity, moving to the small village of Hulshorst, east of Leiden.

Though effectively hiding from the Nazis, Oort did learn of Hendrik C. van de Hulst's work on the interaction of gas and dust in space. Oort then asked van de Hulst to review the available data on cosmic static and to see if he could identify the existence of electromagnetic spectral lines in radio waves. Since there were chemical fingerprints, both absorption and emission lines, in the visible region of the electromagnetic spectrum, Oort reasoned that there could be ones in radio wavelengths, and if so, they could be used to study interstellar gas in the galaxy. van de Hulst ultimately argued that there should be a spectral line marking the change in the energy state of neutral hydrogen atoms. That's what astronomers now call the 21-centimeter line. It gets its name from what happens when the atoms change their state. They emit a particular wavelength of radio energy at 1,420.4 MHz, corresponding to a wavelength of 21 centimeters.

Though van de Hulst calculated where the line should be and reasoned it should exist because there was an abundance of hydrogen in the universe, he still wasn't sure if it would be an emission or absorption line. Still, Oort was awestruck; he could now clearly see the promise of radio astronomy and grew determined to build radio antennae in the Netherlands, an endeavor that would eventually launch an entire research effort to study the radio waves of galaxies and the movement of gas within them. Like Gamow's big ideas about the cosmos, Oort's work would also captivate Vera and entice her to study galaxy dynamics, and later lead to the collection of data on galactic gas that would bolster her own optical telescope data on stars and hot gas moving too fast around the core of galaxies.

That would be years down the road, though. First, Vera would take on Gamow's question of the rotating universe, and she'd dive into it not long after she and Bob were married.

Their wedding, however, didn't go quite according to plan. They had been slated to get married August 22, 1948. Weeks before the date they'd set, Vera and Bob had begun looking for an apartment to move into after they were married. In early summer, one of Vera's friends called and said her aunt, who was roughly equivalent to the general manager of an apartment complex in DC, was going away for the summer, and Vera and Bob could have her apartment if they wanted it. That evening in June, Vera left the house to call Bob. She didn't want her parents to hear her ask Bob if he wanted to get married earlier, so they could live together while they worked in Washington for the summer. Nervously, Vera asked him. Bob said yes.

Excited and triumphant, Vera walked home and calmly told her parents she and Bob wanted to move up their wedding date. Her parents didn't object, though Vera did overhear her father say to her mother: "Everyone will say she had to get married." Vera didn't care. She called the Naval Research Lab, where she was going to work that summer, to ask if she could she start a week late. At first, the answer was no. But when Vera explained she was getting married, she was granted an extra week off to plan the wedding. Bob, who was working at the Bureau of Standards, wasn't so lucky. In fact, the day of the wedding, June 25, he got up, got dressed, and went to work. The couple and their family and friends met downtown later that day at the Statler Hotel at Sixteenth Street and K, and Bob and Vera exchanged vows and rings. They stayed the weekend at the hotel, using a $75 physics prize Vera had won from Vassar. "That paid the hotel bill," she said, "and then we went back to work on Monday morning."[13]

The couple spent the summer in DC, and in the fall, they were off to Cornell, where Vera would begin her master's degree in astronomy.

Figure 4.1
Vera Cooper, nineteen years old, married Robert Rubin in downtown Washington, DC.
Credit: Rubin family.

Her first day there didn't go well. The man who would become her adviser, William Shaw, met her at the door and told her, "Go find something else to study. . . . There [are] no jobs in astronomy; there [are] very few observatories; nobody needed more astronomers."[14]

Vera ignored Shaw's advice. She went on to do a master's thesis, sparked by her fascination with Gamow's question on whether the universe rotated. Her conclusion was shocking—so shocking that Shaw ultimately offered to present her data at a national astronomy meeting under his own name.

Vera said no. She'd present her work herself.

5   Chasing Stars

Though Vera had a shaky start to her first day of graduate school and was told to find something else to study, she'd hoped the discouraging words from her adviser, William Shaw, were not to be taken seriously. Yet as the weeks passed, she was unassured. Tensions between the two didn't improve, leaving her to rely heavily on her husband to help her find her way. It was Bob who pointed Vera to George Gamow's 1946 speculative paper on the rotation of the universe. And it was Martha Stahr, an expert in galaxy dynamics, who helped her collect data on galaxies' velocities, which she thought might help her answer Gamow's question. Tackling that question would inevitably establish Vera's professional relationship with Gamow, and that relationship would become essential to her continuing her career in astronomy.

First, though, Vera had to navigate earning her master's degree in a program with few astronomers and, as a result, few astronomy instructors. She was employed as a graduate assistant in Shaw's elementary astronomy course, which unfortunately conflicted with the usual first-year graduate physics class that Vera was supposed to take. Because of the schedule conflict, Vera wouldn't be able to follow the sequential set of courses in classical mechanics and classical electrodynamics her first year. Fortunately, Richard Feynman, Vera's secret crush only a few years earlier, decided to help her. He was Vera's physics adviser (Shaw was her astronomy adviser) and talked her into taking his quantum electrodynamics course before taking classical electrodynamics to avoid the conflict with Shaw. After taking Feynman's class, she studied with Phillip Morrison and Hans Bethe. "What you must say in all honesty for Cornell was that its physics was very strong," she noted.[1]

Vera did get a peek at how astronomy was done in the summer of 1949 when her parents offered to drive her and Bob cross-country and stop at the world-renowned Lowell Observatory in Flagstaff, Arizona. The family took a tour of the telescope, which was brief, and there was a quick discussion of the science being done there, which delighted Vera. But it also left her with a twinge of envy. "These young men were already doing astronomy," she recalled, "and I was hardly beginning."[2] She felt a pang of despair. Men her age were already at the telescopes observing, and she couldn't even get her thesis adviser to take her interest in astronomy seriously.

When she returned to Cornell and another semester passed, she finally started to take more astronomy classes and had some encouragement for interest in the subject from Martha Stahr. Stahr had recently come to Cornell after teaching at Wellesley and studying at the University of California, Berkeley, where she had finished her master's in statistics in 1943 and her PhD in astronomy in 1945. From 1944 until 1945, she had also worked at Lick Observatory. Most of the astronomers had left during the war, Stahr explained, so she went there to help run the observatory, though there was discussion of whether a woman could handle the telescope. The observatory had a 36-inch reflecting telescope called the Crossley, a moderately ungainly Newtonian telescope, and the man running it "was doing all he could to handle it," Stahr said. "It wasn't long [before] I was doing it with him. . . . It was more than one person could handle."[3] Stahr tended the telescope and also used it to measure the velocities of fifty stars that at the time were considered far from the plane of the Milky Way galaxy. She wanted to show that the velocities of the stars could be recorded to study the galaxy's structure.[3]

Stahr joined Cornell's astronomy department in 1950 when Vera was well into her master's course work and taught a course in galaxy dynamics. Vera found the movements of the stars in galaxies and the movement of entire galaxies captivating, and the course piqued her curiosity further. It "really set me off on a direction that I followed almost my entire career," she said. While in the class, Vera began to seriously consider her thesis topic and finally locked onto Gamow's question of a universal rotation in the universe.[4] Vera decided she would investigate the question because, after all, it made some sense. Planets circle a central point, and so do stars in galaxies. Why not the entire cosmos too?

Evidence was starting to emerge to suggest that such large-scale motions existed, with hints coming from Harlow Shapley and Adelaide Ames's catalog of galaxies. First published in 1932, it revealed galaxies clumped in what looked like superclusters—conglomerations much larger than clusters such as Coma and Virgo. Still, some scientists assumed that galaxies were randomly strewn about all of space and uniformly receded away from us as the universe expands outward. Vera, however, was never convinced by assumptions. She didn't like them, mainly because accepting them without question seemed unscientific. So she decided to see if she could find evidence of large groups of galaxies moving together.

"The only motivation that I can point to is just plain old curiosity," she said. "It was just a question that seemed worth answering."[5]

Having no access to large telescopes to gather her own data, she started to write to observers at other institutions asking for their measurements of galaxy velocities. Stahr helped her with this, contacting Mount Wilson astronomer Milton Humason, who had just the data Vera needed to begin to investigate whether the universe rotated. Stahr explained to Humason what Vera wanted to do and asked him to send her his galaxy data. No, he said. He would be publishing his results soon, and Vera would have to wait until after his measurements were circulated in the scientific literature. That was a letdown for her, and there were others too. There were whispers that mathematician Kurt Gödel at Princeton was also working on rotating universes, and therefore Vera should wait before doing her analysis. "From both the observers and the theorists that were in contact, there were reasons why I shouldn't do this," she recalled.[6] But she didn't listen. She used data on galaxies' velocities, specifically how fast they moved toward or away from Earth, and the galaxies' brightnesses, which were published in the *Handbuch der Astrophysik* by Lick Observatory astronomer H. D. Curtis, and she drew on the theory Jan Oort helped to develop in the late 1920s, which showed that a galaxy's rotation was driven by the way its stars circle around its center. "It just seemed to be a very logical extension to take the Oort theory of galactic rotation as it applied to stars and galaxies and just see if you could apply it to galaxies in the universe," she explained.[7]

By this time, early 1950, she was pregnant with her first child yet adamant about answering Gamow's question, so she gleaned all of the information she could on a set of thirty-eight galaxies lying relatively close to the Milky Way. She tracked the way they moved and looked at the story the data told her. Then she tried again with 70 more galaxies. Collectively, the 108 swirls of stars did seem to exhibit some extra, unexplained sideways motion. When she showed Shaw her work, he said two things: first, *data* is a plural word (Vera had written "data is" in the report), and second, "he made another comment, which I have never repeated, to the effect that the work was pretty sloppy," she recalled. "That, he said, was because I learned things fast, and I didn't worry enough about the details."

Despite his criticisms, Shaw said he thought the results should be discussed at the 1950 winter meeting of the American Astronomical Society (AAS), the preeminent organization of professional astronomers in North America. Given that Vera's baby was due not long before the meeting and that she wasn't a member of the society, he suggested he present her results for her . . . under his own name. It was a blatant example of sexism, and Vera did what she almost always would do when faced with the expectations of gender and academic rank: she defied them.

"Oh, I can go," Vera told Shaw.[8]

David Rubin, Vera and Bob's first child, was born on November 28, 1950. Roughly three weeks later, she was on her way to the winter AAS meeting with Bob, David, and her parents. Vera and Bob didn't have a car at the time, and Vera didn't know how to drive, so her parents drove to Ithaca from DC, picked up Vera, Robert, and their new grandbaby, and then headed south again. The trip was harrowing. It was snowing, and the route from Ithaca to Haverford, which is close to Philadelphia, went directly over the Pocono mountains. Vera's father, who was driving, would later tell her that the trip aged him twenty years.

It was a long, white-knuckle ride, but the family finally made it safely to Philadelphia and were met by Vera's mother's and father's families, who still lived in the area. The day before the talk, Vera's mother and aunt took her to buy a new dress. Then she returned to where the family was staying and rehearsed her presentation. "I had prepared the talk enormously carefully, as I used to. I essentially memorized a ten-minute talk, every sentence," she said. She got up the next day, December 30, went to the conference, and meticulously presented her data and discussed her conclusion. There was, she argued, some kind of extra, collective motion among the galaxies she studied. The data suggested there could be some universal rotation in the universe. Everyone in the room scoffed at her claim. Only one prominent astrophysicist, Martin Schwarzschild of Princeton, responded in a somewhat encouraging way. But even he gently pointed out that to conclude with certainty that groups of galaxies exhibited any extra, collective motion, some sort of collective rotation, she would need more data. "This is very interesting," Vera recalled him saying, "and when there are more data, we will know more."[9]

The press was a bit more liberal in reporting on Vera's results. The story, written by Associated Press science editor Howard Blakeslee, explained that a "young mother, in her early 20s, startled the American Astronomical Society" with her report that the group of galaxies Vera studied was 35 million light-years from the Milky Way, "three times farther away than the greatest telescope can see," and, he wrote, her calculations suggested she'd found "Creation's Center."[10]

The press coverage was sensational. But the reaction of astronomers in the room after her talk was less than lukewarm. Still, Vera did publish an abstract of her research in the *Astrophysical Journal*. The editor at the time, Yerkes Observatory astronomer William Wilson Morgan, told her that she could not title the article "Rotation of the Universe,"[11] but rather "Differential Rotation of the Inner Metagalaxy", which suited Vera just fine.[12] Even with the grumbling in the room after her talk, it never really registered with Vera that her work had unsettled so many astronomers: "The truth is I was so engaged with this child, and the mechanism of getting [to the talk], and I had a new dress, and I thought I gave my paper as well as I could have, so I walked out quite satisfied," she recalled. "I was certainly mildly distressed by some of the comments although I guess I thought that's how astronomers behaved."[13] Schwarzschild's critique, however, she took to heart. She too wanted more data.

Without access to a telescope powerful enough to collect the data she needed and refusal of others to share their observations, Vera was stuck, and with the talk done and the abstract of her work published, she felt lost. She had completed her master's oral exams in October, David was born in November, and she presented at AAS a month later. Now, she was waiting for Bob to finish his PhD, which he was slated to do in June. She didn't have a job or even job interviews lined up. She was just waiting. "In all honesty I had no idea where I was going next. I didn't know what was going to happen," she said, and so she wondered: "Will I ever really be an astronomer?"

Bob was aware of his wife's angst—so aware, in fact, that as he began hunting for his first job, he made sure to choose one located in a city that would allow Vera, at the very least, to continue her education if she wanted to. He'd been offered a job at Oak Ridge National Lab in Tennessee, but didn't think there'd be enough opportunities there for Vera to have a chance at pursuing a PhD in astronomy. He turned that job down and several others. He ultimately accepted a position at the Johns Hopkins

Applied Physics Lab (APL), which was close to DC and had more options for Vera to continue her studies, if she wanted to.

Bob "was my closest professional colleague, as well as my closest emotional supporter," Vera recalled. "He knew from the very beginning how enormously important it was to me to do astronomy, and he's an enormously kind person. He just would not have thought of doing anything that was not helpful to my career."[14]

And so the Rubins prepared to move from Ithaca back to the DC area and for Bob to start at APL. One of the draws of the lab was that Ralph Alpher worked there, and so did physicist Robert Herman, Gamow's collaborators. "Knowing that it was APL and those people were involved made it much more appealing to Bob," Vera said. He was particularly interested in working with Herman, who, with Alpher, in 1948 hypothesized that the residual radiation of the big bang would have a temperature of roughly 5 degrees Kelvin.[15] Herman was also investigating the absorption of light by molecules—molecular spectroscopy—and that interested Bob, who eventually ended up sharing an office with Alpher. The families became fast friends.

Though Bob had picked his job to help Vera continue to study astronomy, the couple had decided that Vera wouldn't start on her PhD right away. Bob and Vera were planning to have another child, and with one less than a year old and not yet certain where they would live, they didn't think it was practical for Vera to start taking classes. Her priority when she, Bob, and their son David moved to DC in 1951 was "to find a place to live," she said. "I really came from a very conventional, very loving Jewish family, and one had a nice home, and everything was done well," she said. "The idea of going off to school and not furnishing a house and so forth just seemed so complicated that I just took time and made a house."

The Rubins moved into an apartment in the suburbs east of Silver Spring. Vera set up the home as she intended, only to quickly realize she was not cut out for suburban living. She didn't have much in common with the other mothers of young children living nearby, and few, if any, would have understood why she wept every time the *Astrophysical Journal* arrived at the Rubins' home. Vera had subscribed to the journal to stay current with the research that was being published while she was out of school, but reading the articles tugged at something inside her, making it all too clear to her that she missed doing astronomy. One particular article, she said, on the masses of double galaxies made her recognize just how dissatisfied she was not doing astronomy. She was not "dreadfully unhappy" during the six months she stayed home with David, but that time probably was the gloomiest of her life, she recalled. "There was nothing in my background that had led me to expect that [Bob] would go off to work each day doing what he loved to do, and I would stay home," she said. "It was just too tough not doing astronomy."[16]

Bob finally told Vera to go back to school. She agreed she should, and yet another concern popped into her mind: telling her mother. Vera worried her mother would think she wasn't "being a very good mother," she said. "I wasn't sure how she would take the news that I wanted to go back to school." Surprisingly, her mother fully supported Vera, offering to help if she could. "She did all these things for all these people, and she said very directly that she would just as soon spend some of her energies helping me as doing these other things."[17]

Another person to encourage Vera to go back to school was the legendary George Gamow. About a month after the Rubins had moved to Washington for Bob's job at APL, Gamow called Vera. He had heard about her talk in Haverford while he was visiting

Princeton and talking with Martin Schwarzschild. Gamow now explained that he was giving a talk at APL and wanted the details of her study to discuss with his colleagues at the lab. Vera eagerly shared her work with Gamow, even though she would not be allowed to attend his talk. APL had a rule at the time that no spouses of employees were allowed beyond the lobby.

Though she was not allowed to attend Gamow's talk discussing her work at APL, his phone calls and correspondence once again tied Vera into the astronomy research of the time. Invigorated by the connection, she started looking at the course catalog at George Washington University, where Gamow was on the faculty. It didn't have a graduate program in astronomy, only physics. Vera was crestfallen. She wanted to put her mind to astronomical problems, not physics ones, she said, so she researched the astronomy program at Georgetown University and talked with the director of the astronomy program, Father Francis Heyden. "I was pregnant, I was Jewish, this was a Catholic boys' school, and I told him I wanted to enter the graduate school, having had a master's. And he said yes," Vera recalled. "If he had said no, I don't know what I would have done."[18]

Vera started working on her PhD in February 1952.

Tuition wasn't cheap. To pay for continuing her education, she applied for a fellowship from the American Association of University Women, but the organization wouldn't fund her graduate studies. "We won't support you because you're going to finish whether we support you or not," Vera recalled the woman interviewing her for the award telling her. "We have to support someone who wouldn't finish unless we supported them." To pay Vera's tuition, the couple dipped deeply into Bob's salary, leaving little money left over.

But perhaps more challenging than money was the logistics of getting Vera to class twice a week. She still hadn't learned to

drive, and Vera and Bob had only one car at the time. Going
back to school became a family affair, with Bob leaving his
APL office in Silver Spring at 5:00 p.m. He would drive to
Vera's parents' house and pick up Vera's mother, then drive to
the Rubins' house, where Vera had fed David and gotten him
settled for her mother to watch him. Then Vera would get in
the car with Bob, and he would drive her to Georgetown for
her classes, which were all in the evening. (That was by design;
Father Heyden wanted to accommodate students who worked
or had other constraints, such as growing families.)

Vera's classes were in Georgetown's observatory, so Bob
would drop her off and sit in the parking lot and wait. Sometimes he'd eat dinner, then come in and work in the library.
The courses lasted two or three hours, and afterward the couple would head home.

The difficult schedule necessitated that Vera move through
her schooling quickly, and Father Heyden obliged, helping her
to plan to finish her PhD in only two years. While her time at
Cornell had been challenging, it helped her complete her PhD
quickly. She had taken "an enormous amount" of physics, and
so for her PhD, she had no physics requirements to finish. And
thanks to the fair number of philosophy classes she'd taken at
Vassar, she had no philosophy course requirements to complete either. "I had gotten very interested in the philosophy
of science [there], so I satisfied all of those requirements," she
explained. "And Georgetown accepted that in lieu of their PhD
requirement for philosophy." Father Heyden, Vera noted, was
"every bit" responsible for smoothing her way with the course
requirements, acting quite the opposite of Shaw, her master's
thesis adviser. This was how Heyden operated, Vera said. He
"collected around him strange people who needed special care,
and he got . . . great pleasure out of making it work for us."[19]

As she plowed through her course work, Vera also started doing small research projects in search of a PhD project. She used the observatory's 5-inch telescope to try to measure stellar radial velocities—the speeds of stars moving toward or away from Earth. It was tedious work, but these were vital data for studying galaxy dynamics. Still, this work didn't capture Vera's attention the way the rotating universe question had. She then worked on identifying faint, unknown lines in the spectra of the sun, a project one of her professors, Carl Kiess of the National Bureau of Standards, suggested she expand into a thesis.

Vera wasn't very interested in that project either, and again she turned to Gamow for inspiration.

6 A Clumpy Cosmos

In the spring of 1952, Vera, pregnant with her second child, climbed into the car with Bob to drive to the Carnegie Institution's Department of Terrestrial Magnetism. Founded in 1904, the research department, nestled on the west side of Washington, DC's famous Rock Creek Park, was originally focused on mapping the Earth's geomagnetic field. With time, though, DTM's scientific agenda expanded to include a more detailed exploration of our planet and also its place in the cosmos.

When Vera first visited DTM, it was brimming with scientific vitality. She was headed there to meet George Gamow in person and discuss a project that she might pursue for her PhD thesis. She had been talking with Gamow occasionally on the phone over the previous months, and now she was thinking of asking him to be her thesis adviser. Dressed smartly in a brown and white striped seersucker skirt and jacket—on loan from Ralph Alpher's wife, Louise, now a close friend—Vera walked into the research institute, followed a corridor filled with books to the library,[1] and within minutes knew this was a place she wanted to work, if they'd ever hire her.

As she opened the library door, there was Gamow waiting for her.

Though he was a brilliant scientist, he "was childish in his enthusiasm for puzzles, games, and scientific tricks," Vera recalled. "He generally carried a deck of cards and would start a lecture by attempting to pile the deck, one-by-one, successively farther over the edge of the lecture table. From another pocket would come two small balls connected by 20-centimeter string. He would start the balls rotating and watch the constraints imposed."

He was also possessed a membership card to a "space cadet corps," obtained by sending off coupons from cereal boxes.

"Gamow could not spell; he could not do simple arithmetic," Vera once wrote. "But he had a mind that allowed him to understand the universe."[2]

Gamow had been intrigued with Vera's master's thesis results suggesting galaxies exhibited some extra, unexplained sideways motion, a hint that the swirls of stars had large, collective movements; it was an idea that fit with his vision of the universe as a vast expanse filled with massive groups of galaxies, possibly orbiting a central point. Meeting in the library, Gamow most likely discussed these ideas with Vera, along with her master's thesis, but exactly what they talked about isn't clear. She couldn't recall their conversation. All she could remember was the seersucker suit she wore and that talking with him "tied me in a little bit with what he was doing in cosmology."

Gamow eventually asked Vera to work on a tricky question: Is there a scale length in the distribution of galaxies? In other words, are galaxies spread relatively evenly throughout the universe, or are they clumped together and moving in structures even larger than the galaxy clusters? There'd been hints of these large structures and motions. Gamow wanted to know for sure if they existed, so he put that challenge to find out to Vera. He gave no direction, no guidance, just posed the question itself. "Everything else," she recalled, "I had to imagine myself."[3]

Vera looked to the literature for some answers and a starting point. Clues that galaxies could clump together in groups much larger than clusters came from photographic plates taken at Lick Observatory, which showed lines of galaxies spreading through space in long strings. Vera's master's thesis, which argued there were groups of galaxies rotating in different clumps around the universe, had also suggested this kind of clumping in superclusters. But she'd been told before that she needed more data to support her conclusion of collective galactic motion, so recalling that criticism as she got to work on Gamow's question, which she chose to adopt for her PhD thesis, she went looking for the best data sets she could find.

Meanwhile, she got a note from Martin Schwarzschild at Princeton, the only astronomer who had given some encouragement to her master's thesis work. Gamow had requested that Vera send Schwarzschild the full manuscript she'd written on her master's thesis to get his thoughts on publishing the entire paper rather than only the abstract. Schwarzschild read the manuscript, then encouraged her to push forward on her PhD thesis rather than spend any more time on the master's thesis work. "Certainly I would think the main purpose of your rotation work must have been to pointedly draw the attention of astronomers to this important problem and to make yourself a first stab at it," he wrote. "Both you have successfully done by your presentation at the Haverford meeting and by the publication of the abstract."

Moving forward on the project wasn't worth it, Schwarzschild explained in a letter to Vera, because new velocity data would soon come out from Mount Wilson, because mathematician Kurt Gödel was continuing to work on his theory of the rotating universe, and because Vera's technique would have to be "somewhat sharpened up" before becoming "sufficiently secure." Still, he told her not to be discouraged or surprised at other astronomers' critiques. "The topic you had selected for your first work is exceedingly fascinating . . . but it is also exceedingly tricky." That said, Schwarzschild noted that he was intrigued with Vera's current question, which Gamow had shared with him, about the clumpiness in the universe. "I sure must admit that you are sliding from one fascinating topic into the next." He urged her to share her results when she had completed her analysis and sent his best wishes for her "big event in the next few weeks."[4]

That big event was the birth of Vera and Bob's second child. The couple had started preparing for the arrival of their baby by spending weekends hunting for the perfect home. On a Monday in September 1952, after a weekend of house hunting, Vera woke up at 2:00 a.m. in labor. In the hospital parking lot, Vera and Bob met her parents, who were going to watch David while Vera gave birth. She left David with them and went inside the hospital and waited. And waited. And waited. "The hospital was very busy, and I was put into a hall, totally neglected. Every time I tried to get a nurse, one would say, 'Can't you see we're busy?'" Vera recalled.[5]

When a nurse finally checked on Vera, it was 4:50 a.m., and now she was nearly ready to deliver her baby. Before Vera could protest, the nurse gave her a shot that knocked her out. Twenty minutes later, her baby girl, Judith, was born; Vera had been sedated and missed the delivery. "I was very angry—I had gone through the entire labor totally alone without a word from anyone and then missed the delivery." Still, her baby was healthy and had not been slow coming into the world, so Vera had no other complaints.

There was little time for celebration. Eight days after Judith was born, Vera went to Georgetown to register for her fall semester of classes (her father stood in line for her until it was her turn), and soon she and Bob and Vera's parents again began their elaborate routine to ensure Vera could attend her classes, raise a family, and continue to wrestle with Gamow's question.

Gamow was after just a single number: the average distance between galaxies. But he left vague the way to answer his question since it was based on sampling galactic distances and was impressionistic. Vera had to find her own way. "That's the kind of person [Gamow] was," Vera said. "I knew what I was getting into. He just threw out lots of interesting questions." She knew she needed data on galaxies, and so, with the help of Father Heyden, she tried to obtain galaxy counts, data from a famous catalog of galaxies, from Harvard College Observatory. Heyden had worked on his PhD at Harvard during the war and had contacts, such as Harlow Shapley, head of the observatory there, from whom he could get galaxy catalogs.[6] Once Vera had those data, however, she wasn't quite sure what to do with them. She had to start from scratch, teaching herself how to approach the problem. It wasn't easy to find people to help her. Unlike Princeton, where researchers and students often discussed their work formally in colloquia and informally over coffee, Georgetown's academic dynamics were different, more flexible to meet the needs of students like Vera. But guidance on research questions was limited. Vera's husband offered encouragement and clues about the method to use, and through him she found another ally, François Frenkiel, a hydrodynamicist and "outstanding mathematician" at APL. Vera met Frenkiel "once or twice in these little cubby holes" at APL because wives weren't permitted there, and "then he finally decided he wasn't going to put up with that anymore, and he got me permission to go to the lab," she recalled. "So after that, I would see him in his office."[7]

Frenkiel gave Vera invaluable advice. He suggested she use a specific mathematical tool, a two-point correlation function, to assess the distribution of galaxies in the Harvard data. Astronomers had already developed a rough idea of how much distance should be between each galaxy if the galaxies were spread evenly throughout space. From that rough distance calculation, he said, she could devise a probability of hitting on a galaxy based on moving some distance from a starting galaxy and apply those probabilities to actual data. From there, Vera could calculate the probability of finding a galaxy at a given distance from a particular galaxy in the Harvard galaxy counts. If the calculated probability, based on the data, was higher than the probability expected based on the assumption that galaxies were randomly distributed homogeneously in the universe, then actual galaxies were more bunched up than they should be, possibly because of violent movement, or turbulence, in the medium that made up the early universe.

"It became clear that one way of doing this problem was to take these counts of galaxies along the sky and apply the two-point correlation procedure," Vera said. And so she sat night after night, when she was not in class, calculating the probabilities of finding galaxies in regions throughout the sky. It was tedious work, done by hand, most often at her dining room table.

Periodically she would meet with Gamow, at either the library at the Department of Terrestrial Magnetism or his house in nearby Chevy Chase, Maryland, to discuss her progress. "It was sort of pleasant. I would tell him what I was doing, and he would tell me what he was doing," she said, though not all their get-togethers went so smoothly. "I remember one or two meetings in his house that were embarrassing to me. I guess he was in the process of breaking up with his wife," Vera recalled. "He would always accuse [her] of something; he would always scream to her. I don't think I ever saw her. I didn't know if she was even in the house." Gamow assumed she was and would scream to her: Why had she gone through his papers? Why had she taken certain papers? Why couldn't he find what he wanted?[8]

Despite Vera's uneasiness in response to these outbursts, she continued to work with Gamow and pursue his question of the universe's galactic distribution. As she analyzed the Harvard galaxy counts, she noticed something odd: inconsistencies in measurements of how bright the galaxies were. The brightness of an object offers a clue to how far away it is from Earth, and so Vera needed accurate measurements of the brightness of galaxies, and therefore distance, to ensure she positioned the galaxies she studied in the correct region of the universe.

Noticing the discrepancies, she wrote to Shapley, who led the project to collect data on the galaxies, to see if he could help her figure out the problem. Shapley replied, saying, "I hear you think there's an error in the Harvard galaxy counts," Vera recalled and asked for details. She and Shapley then corresponded back and forth to see if they could identify the problem and fix it if it existed.

As Vera labored with her analysis of galaxies, Gérard de Vaucouleurs, an astronomer working in Australia, published a paper in the *Astrophysical Journal* suggesting that the Milky Way and its neighbors were situated in a massive clump of galaxies called a supergalaxy. This discovery, he wrote, offered evidence, independent of Vera's master's thesis and the galaxy catalog, showing long, thin wisps of galaxies, that such swirls of stars did bunch together in structures larger than clusters— answering the exact question Gamow had asked Vera to investigate.[9] Without knowing it, Vera had inserted herself into a contentious facet of extragalactic astronomy: Was there large-scale structure in the universe? De Vaucouleurs suggested there was. Her PhD thesis could corroborate the claim or overturn it, and key to answering the question was the Harvard galaxy counts.

Vera got a chance to probe more deeply into the potential errors in the data at a summer conference in 1953. It was a four-week summer school to which she had applied to the National Science Foundation for funding to attend but was turned down. She decided she would go anyway, for two weeks at least. Bob came with her, and her parents watched her children, now ages three and almost one. It was an experience that validated her career choice. "The Michigan summer school was the summer that I really got to meet astronomers," she said.[10] Gamow was there, along with Walter Baade from Mount Wilson, Geoffrey and Margaret Burbidge from the University of London, and many others. Gamow was Vera's advocate. He made the summer school "very nice for me," she said. "He arranged one or two long afternoons to talk to Baade. And we had a couple of long afternoons talking about galaxies." The group also spent a lot of time discussing stars, mainly because a freshly minted PhD student from Caltech, Allan Sandage, also attended the summer school and gave a lecture based on a lot of data he had collected on clusters of stars. In his analysis, he'd compared the stars' intrinsic brightnesses to their colors on what is called a Hertzsprung-Russell diagram, a powerful interpretive graphical tool that astronomers used to illustrate how stars age over time.

The diagram is also useful for estimating the distances to star clusters and for illustrating the enormous physical differences among the stars, such as the existence of giant suns among the far more common dwarfs, of which our sun is typical.

Dwarf stars fall on a curvy line from the top left to the bottom right, called the "main sequence," and the giants are in the upper right. Astronomers thought for decades that the giants were young and through gravitational contraction became the dwarfs. But there was a puzzle that astronomers couldn't quite solve: Why did different star clusters seem to follow different trajectories and therefore have different Hertzsprung-Russell diagrams? Most of these clusters had stars on the main stellar trajectory, called the main sequence, but there were also stars that deviated from that sequence at different points and moved into the giant star region. This was hard to explain by the contraction theory, and of course Gamow thought he knew why. He had been studying stellar evolution and the synthesis of helium from hydrogen within the stars,[11] and so he famously said to Walter Baade: Tell me where the stars leave the main sequence, and I will tell you the age of the stellar clusters.

Following suggestions by others, Gamow assumed age caused the exhaustion of hydrogen in the cores of stars that had powered them for eons, and this was the factor that shifted them off the main sequence. Massive stars shifted more quickly than less massive stars, so it was a fair guess. Just how this happened, however, remained controversial for years, though Gamow turned out to be right. His theory did require significant adjustment by others, such as Martin Schwarzschild. Nevertheless, Gamow was well known for seeing creative solutions to cosmic problems long before there were data to support them. His "intuition was just unbelievable," Vera said. "But he was also fairly unconventional. He did not do the nitty-gritty detailed work." He just could imagine the solution, which "was very inspirational, just seeing a mind work like that."[12]

Hearing Sandage present his initial observational study of the star diagrams in 1953 at summer school and tracking Gamow's contentious debates with other astronomers, Vera realized that stars weren't the only thing astronomers were arguing about—controversy, was rampant in the field, she concluded.

While much of the time at the conference Vera listened and tracked the controversies of the time, she was able to talk with Baade, Gamow, and others about inconsistencies in the Harvard galaxy counts. Stellar brightnesses and distances weren't the only thing in question at the time; the brightnesses and distances of galaxies were too. Vera's conversations didn't resolve her questions, and she returned to DC without much guidance on how to move forward with her thesis. She spent fall 1953 going to class and working on her two-point correlation function calculations in the afternoons and evenings, most often when her children were asleep. "I would only let them nap at the same time," she said, so she could get an hour or two of work in. Then when they woke up, she'd cook dinner, and as soon as supper was over, she'd return to work. "I did virtually all of the thesis work at home, most of it [from] 7 p.m. to 2 a.m.," after the children went to bed.

Months later, Vera mailed an early draft of her thesis to Gamow, who was visiting the University of California, Berkeley. He wrote back saying the thesis "looks quite nice," though he "could not force [himself] to plunge into the details . . . but I presume it is all right."[13]

While out west, Gamow must have mentioned Vera's work to statisticians Elizabeth Scott and Jerzy Neyman at Berkeley, who, according to a letter from Gamow, said what Vera was doing couldn't be done. Scott, Vera recalled, had come to DC to attend a statistics meeting at the Shoreham Hotel and gave a talk on the same subject as Vera's thesis. Curious to hear what Scott had to say, Vera arranged for someone to watch the children so she could attend Scott's talk, and when Scott was done, Vera gathered the gumption to walk up and introduce herself. "I think she must have known my name, knew that I was working on this," Vera said. She had hoped Scott might talk with her a bit more about her work, but Scott said she didn't have time. It wasn't Vera's first rebuff that year. At the Michigan School, she'd knocked on the door of University of Michigan astronomer Dean McLaughlin and asked him if he had time to talk. He too said no.

The next no she got was from the University of Chicago's Subrahmanyan Chandrasekhar, then the editor of the *Astrophysical Journal* and a giant in the field of astrophysics, renowned for his precision calculations of which stars die in fiery supernova explosions. Vera had completed her PhD thesis and submitted it for publication to him, though Gamow warned her not to send it to that journal. Gamow told her instead to let the National Academy of Sciences publish it. (He may have encouraged Vera to submit the paper elsewhere because he and Chandrasekhar had been warring for years about the inner workings of stars and sensed her work, which he supervised, would be rejected.) Vera ignored Gamow's advice and sent her manuscript to Chandrasekhar, who replied that he would not publish the paper; he had a student working on the same problem and thought he should wait until his student was done, he explained. When Vera wrote to Gamow and told him this, he replied on a postcard: "I TOLD YOU SO."[14]

Heeding Gamow's advice the second time around, Vera sent her PhD thesis to the *Proceedings of the National Academy of Sciences* for publication. It was accepted and published in September 1954 and described how the "space distribution of the galaxies" could be calculated using tools from the physical theory of turbulence, which she argued was physically reasonable if "galaxies have condensed from a turbulent gaseous medium." If Gamow's idea about the beginning of the universe, the big bang, was correct, then galaxies in the early universe formed from infalling gas, which would have violent, chaotic motions, and so, Vera concluded, based on this turbulence theory, the upper limit of the scale length of galaxies should be about 10^7 parsecs. That's roughly 32 million light-years, "of the order of the dimensions of clusters of galaxies."[15] The calculations didn't rule out the existence of supersized galaxy clusters, similar to the superclusters seen in the photographic plates of Shapley and Ames; still, the evidence for superclusters seemed unconvincing and controversial. A month after Vera published her work, Neyman and Scott wrote a paper in the same journal reviewing the results and those of Chandrasekhar's student, D. Nelson Limber. Neyman and Scott reiterated in print what they had argued before in conversations with Gamow: the galaxy distribution question couldn't be solved the way Vera and Limber had done it. There had to be a way to answer the question, they said, just not the way it had been done.

Reading the critique of her work, Vera once again felt defeated and began to question her capabilities. Though she was working on her PhD and investigating cosmological questions few others dared to address, Vera still felt as though she hadn't yet made it within the astronomy community. She hadn't published her work and had it favorably received, as male astronomers her age had done. She'd never been to a leading observatory to study the stars, as they had, because she wasn't allowed. And she had very little guidance in her research, something they had in abundance at places such as Princeton. By comparison, she didn't see herself as a real astronomer and questioned whether she would ever become one. Still, she was not ready to settle.

7 Career Challenges and Galactic Questions

As concerned as she was with whether she'd ever make it as a *real* astronomer, Vera tried not to let the fear of failure weigh too heavily on her mind. "I kept myself cheerful . . . by really telling myself that I was just doing something that very few people had done," she said. "That really kept me going."[1]

The positivity paid off. Vera graduated from Georgetown with her PhD in astronomy in the spring of 1954, and with diploma in hand and her PhD published in the *Proceedings of the National Academy of Sciences*, she started teaching at Montgomery College in Maryland, and then in 1955 accepted a position as a research professor at Georgetown. Her first paper, published in June of that year, was on the structure of the Milky Way. It drew on Jan Oort's work with radio telescopes in which he measured the radial velocities of hydrogen clouds in the galactic plane.

At the end of World War II, Oort had begun forging the scientific and political connections needed to build a large radio telescope to look through the pancake-shaped disk of gas, stars, and dust swirling around the Milky Way's core. He initially pitched the idea of a 25-meter radio telescope, but funding and engineers with the experience to operate such a behemoth antenna didn't exist at the time, so he had to start smaller.

With the help of A. H. de Voogt of the Dutch Post, Telephone, and Telegraph Service, Oort was able to co-opt a single 7.5-meter radio receiver, one of a few he collected from the Dutch coast. The antennae had been cast off by the Germans after the war, and de Voogt was preparing to turn them into research telescopes. He had salvaged the reflectors and brought them to Kootwijk, a Dutch village about seventy miles east of Leiden, and allocated one for galactic studies while reserving the others to investigate Earth's atmosphere and the sun. (As the radio telescope project took shape, Oort suggested a feature of the solar system that he's well known for now: that a cloud of comets, the Oort Cloud, circled the solar system's outer edges.[2]) Almost as soon as Oort and his colleagues got their hands on one of the co-opted radio receivers, they began searching for radio emissions from the Milky Way, specifically the 21-centimeter line—that spectral line corresponding to the wavelength of radio wave emitted when the energy state of a neutral hydrogen atom changes.

As the work ramped up, a fire ravaged the radio receiver, but by the spring of 1951, the team had made the necessary repairs and narrowed in on the long-sought 21-centimeter line, first detected by Harvard researchers Harold Irving Ewen and Edward Purcell. Henk C. van de Hulst, who had predicted the existence of the spectral line, had been visiting Ewen and Purcell in Cambridge, Massachusetts, when the Harvard team informed him that they had detected the 21-centimeter line a few weeks before. They asked van de Hulst if he could write to the researchers at Kootwijk and confirm the result.[3] Van de Hulst quickly scribbled off a letter and shared an essential detail of how to capture the signal. Using this information, the

Dutch team also identified the 21-centimeter line, and both teams reported their results in September 1951. In the Dutch team's paper, the radio astronomers noted that their observations confirmed Oort's earlier results showing the Milky Way did in fact spin about a central point.

Oort went on to use the radio telescope to show that the Milky Way had spiral arms made of stars and gas.[4] Our galaxy was a swirl much like its nearby neighbor, Andromeda. The result spurred another question: How fast did the gas clouds in the spiral arms circle the center of the Milky Way?

Using the Kootwijk receiver, the researchers recorded velocities of neutral hydrogen gas clouds, then plotted the velocities against their distances from the galactic center. They drew the galaxy's rotation curve, extending it as far as the sun. All the gas clouds, the rotation curve revealed, had roughly the same velocities: they circled the galaxy's center at about 200 kilometers per second, speeds consistent as far out as 26,000 light-years from the core of the Milky Way. The galaxy's gas, and its stars, whether close in or far out, seemed to have similar speeds. No one, however, questioned the results or suggested they hinted at the existence of dark matter.

At the time, astronomers assumed that the speeds of stars and gas beyond the galactic orbit of the sun would drop off, following Newton's law of gravitation. Rather than pursuing what the gas and stars did even farther from the galactic center, Oort and colleagues instead used the data to estimate the mass of the galaxy at 70 billion suns[5] and then moved on to the other questions.

Vera was intrigued with the results. She wanted to know whether the gas clouds had any additional velocity, called streaming velocity, or if what you measured was what you got. Using the radio observations (the 21-centimeter-line data), she calculated that the clouds did in fact have a bit of extra streaming velocity—about 5 to 10 kilometers per second more than what Oort originally reported. And she argued that this extra velocity made it necessary to tweak Oort's picture of our galaxy just a tad. Based on her calculations, the gas clouds Oort had looked at were closer to the galactic center than he'd suggested, and that meant one of the galaxy's spiral arms was actually swirling inward toward to the core of the galaxy.[6]

That was Vera's last paper on the galaxy for a while. For the next half decade or so, her research focused on the sun. Part of the funding for her position at Georgetown came from grant money awarded to the school so its researchers could analyze solar eclipse observations, and she helped when she could. By this time, in the fall of 1955, she and her family had moved to the Midwest. Bob had accepted a postdoctoral position at the University of Illinois in Urbana-Champaign, so in 1955 and 1956, the family lived in Illinois while Vera continued her work, remotely, for Georgetown. Once again, Vera felt isolated, she recalled. Urbana-Champaign was a place where she said she felt unwelcome on the university's male-dominated campus. She felt shut out of the university. She couldn't connect with astronomers and physicists nearby. And she wasn't doing the research she would have liked to have been doing. It was once again a trying time, she said.

"I was raising children," she said. (Their third child, Karl, was born in 1956.) "And I was sort of hanging in there, and that was really enough for me," she explained. "I was keeping up with the literature to a degree that I have never since. I read everything that came out. So I was learning," she recalled, just not really doing influential research.[7]

What was influential research at the time was the determination of how the chemical elements in the cosmos were made, or at least how the elements heavier than hydrogen and helium were made. This was research Margaret and Geoffrey Burbidge had started a few years earlier, and as Vera read more about the work and the researchers, Margaret became her role model.

Margaret was born Eleanor Margaret Peachy in 1919 in Davenport, England, to parents who were chemists. Margaret's mother was adamant that both her daughters would have careers, so at her mother's behest, Margaret took her early childhood fascination for numbers and started studying the stars.[8] After earning a PhD in 1943 in astronomy from University College London (UCL), she applied for a Carnegie fellowship, which would have allowed her to use the coveted Mount Wilson Observatory telescopes, but she was denied the award, possibly because she was a woman.[9] She stayed at UCL instead, working with the school observatory's spectrograph, a tool that could be attached to telescopes to record celestial objects' spectra—the amount of light they give off at different wavelengths. In 1948, she became UCL Observatory's assistant director and married Geoffrey Burbidge. She and Geoffrey then worked at Yerkes Observatory where she had an International Astronomical Union fellowship, and when they returned, she started collaborating with British astronomer Fred Hoyle (who coined the term *big bang*), as well as Caltech nuclear physicist William Fowler and her husband on the origin of the elements.

Hoyle had come up with the idea that the elements form by nuclear fusion in the 1940s, and by the mid-1950s, Margaret finally got to observe at Mount Wilson, working as her husband's assistant. (He had been awarded a Carnegie fellowship.) On the mountain, Margaret started using a spectrograph to study stars and the elements within them, while Fowler experimented on the elements in his lab. Collectively, the team looked at the observatory and lab data, framed the data with the theory provided by Hoyle and Geoffrey Burbidge, and mapped out the nuclear processes taking place at the centers of stars. Those nuclear reactions, including the ones taking place in the brief explosive state that more massive stars experience called supernovas, could account for the formation of all the elements from hydrogen to iron and beyond. Their theory accounted for the origin of all of the elements—the ones that make up the stars, the planets, us, and everything else in the universe.[10] It was the work that convinced astronomers that we are made of the material within stars, that we are the stuff of the stars.

That astrophysical theory revolutionized our understanding of the cosmos and how the heavy elements were made. And it did something else: it led the Burbidges to think about the stars in galaxies and how those stars move, which they explored to great depths using a technique that would become essential to Vera's future research.[11]

Before Vera learned that technique, though, she spent a few years staring at spectra of the sun. When she and her family returned to DC in 1956, she was still on Georgetown's payroll and doing even more research on the planets, the sun, and solar eclipses than ever before. She worked directly with Charlotte Moore Sitterly, an astronomer at the National Bureau of Standards, which had many researchers working on atomic spectra at the time. Sitterly was particularly well known for her research on the sun's spectrum. She gave Vera prints of 3-feet-long, 1-foot-high solar spectra from one of the solar telescopes at Mount Wilson and taught her how to measure them—her first formal introduction to measuring spectroscopic data and then interpreting the results.[12]

A year on, Vera received a tantalizing note from Maud Makemson. Though they'd parted on less than stellar terms when Vera graduated from Vassar, Makemson had followed Vera and was impressed with her success. "You are one of the wonders of the world," Makemson wrote, "raising such a darling family, getting degrees and carrying on your interest in astronomy." Makemson's letter, in addition to praising Vera, came with a request: Would Vera consider taking over the directorship of the Vassar College Observatory from her and head the astronomy department? "There is no one whom I would rather see in charge of the department," Makemson wrote. "I wager you would put new life in it!"

She suggested Bob might consider inquiring at IBM, with its "big electronic computers," for a job and urged Vera to respond quickly if she was interested, "before we have engaged a second rate young man," she wrote. She expressed how it would be a "great pity" to hire a man and break the string of women directors who had run the observatory since the observatory's inception in 1865. Makemson was adamant: "The opportunities in astronomy are enormous nowadays," she wrote, hoping Vera would take her up on the offer.[13]

It wasn't the right time to move, so Vera didn't take the position at Vassar. She stayed at Georgetown and continued her work on the sun, starting a project on solar limb darkening—the optical effect seen in images in which the edges of the sun seem much dimmer than the star's center. This dimness appeared in images collected during a total solar eclipse in 1952 from three sites in Africa, and using the data, Vera calculated the brightness of the outer edges of the sun suffering from the limb-darkening effect. Once again, when she went to publish the results, she ran into trouble with Subrahmanyan Chandrasekhar, still the editor of the *Astrophysical Journal*.

She had typed her solar limb darkening paper, one of the few she'd done this way, and sent it off to him for publication.

"It was not embarrassingly typed. I don't send out work that's embarrassingly typed, but I had typed it myself. I thought it was adequate," Vera recalled. Chandrasekhar disagreed. He returned the paper and asked her to retype it and put zeroes in front of all the decimal points. It was an odd request, but she obliged, retyped it with the requested zeroes, and resubmitted it Chandrasekhar. The paper was published in February 1959.[14]

A few months later, Bob received an invitation to participate in a summer school at the Les Houches School of Physics in the French Alps. He replied that he would be delighted to attend and informed the school organizers that he would be bringing his wife and three children. The kids were now eight, six, and three years old, and rather than leave them with their grandparents, the Rubins, with their adventurous spirit, decided they would sail as a family across the Atlantic to France. They rented a small room with bunks and a crib, which made moving around in the space a bit tricky but fun nonetheless, Vera recalled. One evening, the couple left the children to enjoy the evening entertainment on the ship and were stunned to find the new director of Harvard Observatory, Donald Menzel, playing guitar and charming the "not-first-class" passengers, as Vera put it. Menzel, the entertainer, she said, was different from the one she knew from Harvard—a pleasant surprise.

After more than two weeks at sea, the ship docked at Le Havre, and the family drove to Paris and then on to Les Houches. "It was mountain living, and unforgettable," Vera said. "We shared this house with a French student and his wife and two sons, who were the ages of our children. We muddled through with high school French, but the children needed no language to play together."

Bob spent most of his time in class, and a few times Vera and the children would join him for dinner with the other students. "One of the students was an active British astronomer, who dashed around in a handsome red roadster with the top down," Vera said. "His name was Donald Lynden-Bell." He was a charismatic young scientist who would go on, as Vera would, to completely reshape what we know about galaxies.

While Vera tended to the children most of the time, she was invited to attend a class on the evolution of stars taught by French astronomer Évry Schatzman, though it was against the rules of the school since she wasn't the one invited to attend. Breaking the rules and going to the class made the trip extra special, she said.

Before she knew it, though, it was time to head home.

"I was to fly to New York with the children. The plane to New York left about three hours late, midnight instead of 9 p.m.," Vera recalled. "When finally we reached New York, with numerous suitcases, winter coats, and a large duffel bag, the customs officer chose to open the duffel bag. There on top, among much equipment, was a large saw.

"Lady," the customs official said, "how are you going to get out of here?"

Vera pointed to her parents waiting on the other side of the customs line. "He passed us through. He never opened the sleeves of the children's winter coats. All were stuffed full with rocks."[15] (Decades later, two of Vera and Bob's children would become geologists.)

A year later, Vera and Bob went abroad again, this time for Vera to attend an astronomy summer course in the Netherlands. And this time, they left the children behind. David and Karl went to Bob's parents in Florida, while Judy and the newest addition, Allan—born in May 1960—stayed with Vera's parents. It was one of those times, among many others, that the older siblings would take care of the younger ones, most memorably with Judy pushing Allan up and down the street in a stroller. Vera's astronomy truly was a family affair.[16]

On this adventure, the Rubins headed to Nijenrode Castle in Breukelen, where Vera spent three weeks learning about the structure and evolution of galaxies. Jan Oort and Margaret and Geoffrey Burbidge were among the lecturers; Vera, it seemed, was dipping a toe back into research far beyond the solar system and the sun. "Their talks and discussions played a large role in bringing me up to date with current important events in extragalactic astronomy," Vera said. "Initially, Oort terrified me," she reluctantly admitted, "but I soon had too many questions to stay silent."[17] She became fascinated once again by the use of radio astronomy data, revealing the galaxy's spiral structure, which reminded Vera of her 1955 paper on our home galaxy, the Milky Way.

Five years had passed since that work had come out, and astronomers now knew that the galaxy's spiral arms were actually circular, but fairly irregular, with "sudden breaks, crosses, and confused regions," Vera said. They also knew the galaxy was rotating, as Oort had suggested in 1927, yet what still puzzled them was how the galaxy changed with time, how it evolved and how other galaxies did too.

Astronomers knew spirals weren't the only kinds of galaxies in space; the swirls of stars also came shaped as oblong eggs, which raised questions about where each type of galaxy fit on a timeline that mapped out galactic evolution. Astronomer Edwin Hubble at Mount Wilson Observatory suggested that the oblong, elliptical galaxies were precursors of the spirals, but even if that were true, there were other questions that needed answers, Vera noted in a 1960 *Physics Today* article describing what she'd learned at the Nijenrode Castle summer school. One issue was how the spiral structure of a galaxy was maintained, even when stars in a galaxy had differing velocities. Another was whether the stars and gas clumps were moving in a similar fashion. And was gas flowing out of a galaxy into intergalactic space? If so, how was the gas at the center of the galaxy replenished, and did interstellar dust follow the classic rules of chemistry, or was something else going on? There seemed to be an implication that our galaxy also had a gaseous halo that somehow sent matter into the Milky Way's center, but whether that was true was also open-ended.[18]

Our galaxy, and the others in the universe, Vera decided, were ripe for exploration. In attending the summer school, she had found a new research focus. She started corresponding, and then collaborating, with Gérard de Vaucouleurs, on elliptical galaxies. Impressed by Vera's enthusiasm, de Vaucouleurs invited her to visit him and his wife, Antoinette, in Texas. Vera had arranged to go, making plans for the children and for her travel, and then her sister called: Ruth's husband had become deathly ill, and Vera decided she needed to stay in DC to help her and her family. "I was sorry to have to cancel my trip to the University of Texas," she wrote to de Vaucouleurs. "After speaking to you last week, I spent 36 hours with my sister, sitting in a hospital chair. For many hours during the night, there was almost no hope."[19]

Thankfully, her brother-in-law's condition steadily improved, enough for her to visit de Vaucouleurs later that year. And, in turning her attention from the sun and stars back to galaxies, Vera stumbled onto a discussion about galaxy dynamics that had been going on for decades; this one looked at the forces affecting the gravity of galaxy clusters. Vera had finally stumbled on the work of Fritz Zwicky and his stunning conclusion that some form of mysterious matter was cloistering galaxies in the Coma cluster. The idea was bolstered by Sinclair Smith and his look at galaxies in the Virgo cluster, yet not everyone agreed that mysterious matter was needed.

The idea had been lost from the literature for a while, then reemerged in the 1950s when Armenian astronomer V. A. Ambartsumian and his collaborators began studying groups of stars thought to be bound together. A closer look at these stellar clusters revealed that some of the individual stars had way too much energy to stay within stellar clusters for very long; they were being ejected from the system, it seemed.[20]

And so Ambartsumian argued that if this were true for groups of stars, would it not also be true for groups of galaxies, such as the clusters Zwicky and Smith studied? If Ambartsumian was right, it could mean that the theorem Zwicky and Smith had used to calculate the mass of the galaxies wouldn't work—and that would mean there wouldn't be a need for dark matter to keep galaxy clusters together.

Ambartsumian's idea, first published in 1954, got the attention of astronomers, specifically that of Martin Schwarzschild. He was the Princeton astronomer who had responded kindly to Vera's master's thesis presentation, and, prompted by Ambartsumian's analysis of the clusters, he looked at Zwicky's and Smith's papers and then reanalyzed the Coma cluster data. He found that, indeed, the clutch of galaxies had to have much more matter than we could see to keep it together.[21]

Yet Ambartsumian's argument didn't just go away. What he'd pointed out was that there were huge discrepancies in the mass estimates of individual galaxies using the theorem Zwicky and Smith had used for their calculations compared with the galaxy mass calculations made from velocity measurements of stars moving about the center of each individual galaxy.

Ambartsumian's argument that maybe the clusters weren't as tightly bound as astronomers assumed would remove the need for any unnecessary matter. But taking away the unseen matter added another problem: it raised the question of what sent the galaxies flying apart, because they would have to be flying apart if they weren't sticking together. Ambartsumian suggested a mysterious extra energy, maybe from explosions, would do the trick.

The argument for dark matter or some ejective energy sparked a heated debate that continued throughout the 1950s and came to a head just before the International Astronomical Union general assembly meeting in Berkeley, California, in 1961. Vera was slated to attend that meeting, and having been interested in galactic and extragalactic astronomy, she was also invited to attend a premeeting in Santa Barbara where researchers planned to hash out the idea of extra energy or extra matter in galaxy clusters and any implications the conclusions would have for cosmology.

At the gathering, the discussion focused on the different mass estimates for galaxies. "Ambartsumian seeks to explain this discrepancy by assuming excess energy. . . . Alternatively, it can be explained by assuming the presence of extra mass that cannot be seen or detected by other means as yet." This would be mass either in the intergalactic medium, or in very dim stars or other celestial objects, the conference organizers wrote.

Admittedly, the group didn't love the solution that relied on extra matter in the clusters; it implied "that some 99% of the mass of the universe has remained so far invisible, and it is perhaps distasteful for astronomers and cosmologists to think that their theories are based on observations of less than 1% of the matter that is really there!"[22]

And so Ambartsumian, the Burbidges, George Abell, Gerard de Vaucouleurs, and a dozen others launched into discussions of what they called the galactic instability problem, since the galaxy groups were unstable without added, unseen mass. Hours of conversation and deliberation ensued, and still the group couldn't come to a conclusion. Invisible matter was a solution many astronomers were firmly against. But it still couldn't be ruled out, the conference organizers wrote.[23] "It must be admitted," they explained, "that more questions have been raised than settled."

8 A Taste of Astronomy

After sitting through the days of debate in Santa Barbara, where prominent astronomers argued about what was going on in galaxies and galaxy clusters, Vera knew she wanted to contribute to the questions being asked. The group, which met to discuss the need for extra matter in and among galaxies, didn't come up with a conclusive answer on the topics—only more questions. This intense discussion may have prompted Vera to shift her attention from her research on the sun back to galaxy dynamics, a topic she first became fascinated with as a graduate student.

Curious about how galaxies spun, Vera chose to focus how fast stars in the Milky Way swung around the galaxy's center. Having established herself at Georgetown by the early 1960s, she had a bit more leeway to begin working on projects that she had the passion to pursue rather than what was given to her to complete. Drawing on what she'd learned at the 1960 summer course in the Netherlands and at the 1961 meeting in Santa Barbara, Vera began to investigate whether stars and gobs of gas moved similarly around the galaxy's center. Because there weren't other Georgetown faculty to work with—most professors were researchers elsewhere—Vera began collaborating with her students. In their first project, she and her students identified the orbital speeds of galactic stars located less than 10,000 light-years from the sun. The team divided the stars into zones based on their longitude within the galaxy and then compared the stars' velocities with the velocities, taken from radio astronomy observations, of hydrogen gas in the same zones. Plotting the velocities of the stars and gas against the distances of the stars and gas from the galactic center, Vera and her students, including a future NASA astronomer, Jaylee Burley (later Jaylee Mead), drew the rotation curve of the galaxy. Gas and stars, the curves revealed, swung around the Milky Way in much the same way, a finding that raised another question for Vera.[1] Are there data on stars that sit farther from the galactic center than the sun?

She and her students searched endlessly for more information on the Milky Way's stars. It was actually her students who led the way, she recalled; they knew how stars were cataloged, and so she pushed them to identify as many stars as far from the galactic center as possible so the team could draw the rotation curve of the galaxy out beyond the orbit of the sun. Vera and her students worked on this as a class assignment, an exercise that would improve the mass estimate of the galaxy if they were successful. Together, the group dug deep into the astronomy literature looking to find data that would reveal the orbital velocities and galactic distances of hot, blue-white stars. Based on what all astronomers had assumed, the team expected to see the stars move around the galaxy much the way the planets move around the sun, with the closer ones moving faster than the ones farther out.

But that's not what the team found. When Vera and her students drew a new rotation curve using the data they'd collected, they found the velocities seemed to hold steady.

The rotation curve did not drop off. It was flat.

With meticulous attention to detail, Vera and her students wrote up the results and submitted the paper to the *Astrophysical Journal*. Chandrasekhar, still at the helm, once again gave Vera grief about the manuscript. He said he'd publish the abstract of the paper but not with the students' names. Vera responded that if that were the case, she'd withdraw the paper from the journal. Chandrasekhar relented. In June 1962, Vera and her students reported that for distances from the galactic center out to about 27,000 light-years, as far as the data on galactic stars would allow at the time, "the stellar [rotation] curve is flat, and does not decrease as is expected for Keplerian orbits."[2]

Not long after the paper appeared in print, the criticism came rolling in. Astronomers wrote to Vera commenting that the result couldn't be correct or that the data weren't good enough to make such a claim.[3] As she had done before, Vera noted the concerns and continued with her research. This time the negative feedback didn't get to her as badly as it had when she was doing her graduate studies; what did bother her was the need to rely on other astronomers' data to do her work, which left her unable to counter some of the criticism of the results.

Serendipitously for her, the Kitt Peak National Observatory near Tucson, Arizona, had opened a few years before, and it didn't restrict women from observing there, as other facilities, such as the Mount Wilson and Palomar observatories, did. In 1962, Vera applied for time to use the telescopes at Kitt Peak, and by early 1963, she was on her way there for her first official view of the heavens as an astronomer. Inspiring her observational research on the Milky Way was Caltech astronomer Guido Münch, who had been tracking the speeds of stars in the Milky Way and "trying for the first time to deduce a rotation curve of our own galaxy, star by star," Vera recalled.

Münch was looking for stars that he could observe and then used a spectrograph to measure the specific wavelengths of light the stars emitted. He wanted to measure the stars' spectra to trace what he assumed were the stars' orbits around the galaxy's center, but, Vera said, he was focused only on the stars near the inner part of the Milky Way.

Vera wanted to do something slightly different: study the stars in the opposite direction, farther from the center of our galaxy. "I was interested in looking at the outside of the galaxy and trying to deduce the rotation curve for the outer part of the galaxy," she said.[4] Her goal was to obtain the velocities of stars far beyond the sun. Astronomers assumed those velocities would drop off. Vera wanted to check that assumption.

Finally, about a decade after earning her PhD, she would get the chance to collect her own data for the first time. At Kitt Peak, she was slated to use a 36-inch reflecting telescope. Sitting in the observing room, waiting for dark, she was lost in thought when a man walked in, breaking her concentration. She politely introduced herself, to which he retorted sharply: "I know who you are. I was running to get to your talk at the 1950 Haverford meeting of the AAS, and I slipped on the ice and hurt my knee." The man was Art Hoag, a well-known astronomer at the time, and Vera felt her introduction to him hadn't gone so well. She brushed it off and started searching for stars far away from the center of the galaxy and trying to deduce their orbital velocities. "It was an ambitious project," she said, "but I hoped to learn about motions of stars within our galaxy, and motions in more distant galaxies."[5]

As she gathered her data, she suspected she might need help in pursuing her goal, especially if she wanted to study the speeds at which stars in other galaxies whip around their galactic centers. Making those observations were outside of her wheelhouse, so she started thinking about collaborators. Bob had recently learned that he would have a year's leave with salary from the Bureau of Standards, where he was working at the time, as a reward for his research in statistical mechanics. He had gotten an invitation from researchers in Trondheim, Norway, to join them there for his sabbatical, but he thought the Scandinavian winter would be brutal for Vera and their four children. So Bob instead decided to look elsewhere and ultimately settled on working with a physicist in La Jolla, California. That was also where Margaret and Geoffrey Burbidge—who helped establish that most of the chemical elements that compose Earth (and therefore us) are cooked up in the hearts of the biggest stars—were now working, and so both Bob and Vera would have top-notch colleagues with whom to collaborate, if the Burbidges would accept Vera. Bob was in favor of the idea. "He made it possible, he always did," Vera said of her husband's decision.

Vera's next step was to convince the Burbidges that they should collaborate with her, and when she contacted them, she asked to meet with them at the AAS meeting in Phoenix in April. They agreed, with Geoff suggesting they all have lunch together. When Vera walked into the restaurant to meet the couple, another astronomer was there. It was Allan Sandage, once an assistant of Edwin Hubble and the preeminent observational cosmologist of the 1950s and early 1960s. He had, several times, redefined the age of the universe, taking it from 2 billion to 13 billion years old in less than a decade.[6] He was a giant in the astronomy world, and so were the Burbidges, and now, at the AAS meeting in 1963, he was standing there with them, set to join the trio for lunch. Vera immediately realized what was going on: the Burbidges, probably with Sandage's input, were about to interview her and accept or deny her request to work with them for a year. As the astronomers sat down to eat, the discussion moved to galaxies, and Vera almost effortlessly began adding insightful information to the conversation. Key to her success, she said, was that she had read the *Astrophysical Journal* cover to cover since the early 1950s to stay connected with research, even when she couldn't do her own. Vera's research, for the most part, had focused not on galaxies but rather on classifying stars and studying the spectrum of the element iron, along with some analyses of stars' motions in the Milky Way. But because of Vera's close attention to the literature, "there certainly were a few places where I knew things that they did not, because I read the literature so well about galaxies," she said, "so I think I passed that interview."[7]

She did. The Burbidges invited Vera to work with them in the fall in La Jolla. With Bob and Vera's jobs settled for the year, the family decided that to move west, they'd drive cross-country. On a summer's day in August 1963, Bob, Vera, and their four children piled into the car at 4:00 a.m. and started their drive— with their clothes and camping gear strapped to the roof of their station wagon. "En route we camped and hiked and fell in love with Jackson Hole, Wyoming,"[8] Vera recalled. They arrived in La Jolla a few days later and moved into "a gorgeous house overlooking the Pacific . . . it was unbelievable."

Vera immediately began working with the Burbidges, measuring spectra of galaxies that the couple had amassed at their time at the telescope but hadn't had a chance to analyze. Vera's work was often done directly with Margaret, but Vera said she often sat and talked with Geoff: "He like[d] to talk, and he would come in and sit down and talk a lot," she said. He would talk about the galaxies and "grand ideas," and although the Burbidges had contributed so much to astronomy already,

Figure 8.1
Vera sits with her children (left to right) Karl, Dave, Allan, and Judy at Bear Lake in Colorado's Rocky Mountain National Park in 1961.
Credit: Rubin family.

Vera said they never made her feel as though she was their student. They treated her as their colleague. They had better equipment than she had had at Georgetown, but "I certainly had measured enough spectra that I knew how to do it," so the astronomy power couple left Vera to do her work, and if she swiftly wrote the paper about a galaxy's spectrum, they gave her the credit as first author. "Geoff was impatient, but in some sense I am impatient, too. I mean, I would finish measuring something, some object they had studied, and it would be interesting," she explained, and Geoff would tell her to write it up. He would say, "Well, you did all this work. We could write the paper, but why don't you take the weekend, and if you get something written we'll use that. That'll be the start of the paper. And if you don't get around to doing it, I'll do it," she recalled him telling her. "I don't think I was on their level, but they were very willing to let me do the work."[9]

Not far into 1964, Vera put out a bunch of coauthored papers with the Burbidges, and on several she was the lead author, a rarity in her work with collaborators at Georgetown. In addition, the Burbidges let Vera come with them to use the 2.1-meter reflector at the McDonald Observatory in Texas, where she again got a taste of being an astronomer in the truest sense of the word. "The observer stood high at the top of the dome, so I felt as if I was standing on nowhere," she recalled, hinting at how she made the observations from the telescope's primary focal point high in the sky.[10] Geoff usually assisted with the observations, meaning he would carry a tiny one-centimeter-loaded film case up the ladders along the inner edge of the dome, and pass it to Margaret, then climb down with the previously exposed image.

"Loading the film was a little tricky," Vera said. In total darkness, the person carrying the film up had to unfurl a large, skinny roll of it and cut it to the required size, load the cut piece into the holder completely in darkness, return the entire roll back to its dark place, and hope that she had placed the film with the correct side up. "If not, the exposure would be wasted," Vera explained. After a while, Geoff noticed Vera wasn't making any errors and asked how she avoided them. "Easy," she replied. "I cut two pieces, and when the loading is done, I turn on the lights. If the 'tester' has the correct side up, then I know the loaded one is correct. If it is not correct, I turn off the lights, turn over the one in the holder, and I know that it is now correct."[11]

Soon Vera was the one regularly doing the observing at the edge of the telescope dome.

9 At Last, a *Real* Astronomer

By the time Vera was wrapping up her year with the Burbidges, she had analyzed the chemical fingerprints of dozens of galaxies and published details of the rotation and estimated masses of many of them. None, however, gave a hint that galaxies contained invisible matter. That may have been because the Burbidges "always did what everyone did," Vera said: they extrapolated the speeds of the stars beyond where they'd measured using the assumption that the stars' speeds would drop off. At that moment, though, making that assumption didn't bother Vera too much. She had her first taste of what it meant to be a "real" astronomer, and she quickly realized she could become one. Margaret Burbidge had shown her that. "Margaret . . . was studying galaxies. She was married, and she had a child," Vera recalled. It was a life Vera dreamed of living,[1] and she now saw firsthand that it was possible. She could indeed have it all: a career, a loving husband, and a family. As this realization slowly sunk in, in the spring of 1964, the Rubins were preparing to return to Washington. Once again the family packed up the station wagon and slowly made their way east, stopping to camp and hike along the way.

Even before the Rubins had begun their trip back to DC, however, there was a gnawing in Vera's gut that something wouldn't be right when she returned. It specifically had to do with her work at Georgetown. "Starting in '63 . . . just about the same time we went to La Jolla, I started observing at Kitt Peak, and teaching and observing and having four children was just more than I wanted to handle," she recalled. "If it had been impossible to do anything else, I probably would have made it work," but she felt as if something would have to give. Otherwise Vera wouldn't be able to do everything to her level of expectation, which was exceedingly high.

The Rubins weren't home from La Jolla long before Bob and Vera flew to Germany for the 1964 Hamburg meeting of the International Astronomical Union. The children stayed with Vera's parents, while the couple attended the meeting. At the closing banquet, Allan Sandage walked up to Vera as she and Bob were leaving the dance floor. Vera's work had made an impression on Sandage, he explained, and he asked if she'd be interested in using one of the telescopes at Palomar Observatory, even though the facility was not officially open to women. Vera eagerly said yes.[2]

That moment put Vera's problems with her position at Georgetown into perspective.

The astronomy department had very little money—in fact, almost none. As a result, "Father Heyden would take any grant that would give him money, and they were almost always from the Air Force." That meant research could be taken right out of Vera's hands and classified without her ever seeing it again. "And that really wasn't what I was interested in doing." She wanted to study galaxies, and, as she slowly increased her time observing at the telescope to pursue that line of work, Vera realized that continuing to raise children, do research, and teach would be untenable. "It was really the observing that was the last straw, to go away a couple of times a year when I was teaching [and] come back behind."[3]

Vera's family and her observing runs weren't aspects of her life that she was willing to give up, especially when she received a more formal invitation to Palomar in the mail from Sandage in the summer of 1964. It was the application for telescope time, which Sandage had previously promised Vera. Prominent on the form was the warning: "Due to limited facilities, it is not possible to accept applications from women." But Sandage had penciled in the word *usually* above the sentence. Without hesitation, she applied for telescope time and waited to hear if she'd be observing at the renowned California telescopes.

In the meantime, she decided she would resign from her position at Georgetown. The research environment and teaching obligations weren't what she wanted, so she sent a note to the dean of the astronomy department in late December 1964 telling him that she would continue working until June but would not stay beyond the end of the spring semester. She then drove over to the Department of Terrestrial Magnetism (DTM) and asked for a job.

Vera had previously visited the research facility and met with Bernard "Bernie" Burke to talk with him about his work measuring the speeds of gas clouds moving around the Milky Way, Andromeda, and the Triangulum Galaxy, another spiral nearby. She was talking with Burke that day at DTM about galaxies, when suddenly she blurted out her request for a job. "I went home and told Bob that if I had asked [Burke] to marry me I don't think he would have looked more surprised," Vera recalled. "Until the end of the war, even the secretaries [at DTM] . . . were male," she explained. "They had never had a woman staff member."[4]

Burke got over his initial shock of Vera's request relatively quickly. In response, he invited her to join him at the department's weekly staff lunch, where researchers met to eat, talk, and share their work with everyone else at DTM. Just as the staff was preparing to eat, Merle Tuve, the DTM director, came over and told Vera, "Well, we do a lot of learning at lunch . . . and there is a blackboard there, so why don't you get up and tell us what you're doing."

Vera went to the blackboard and started talking about her work on galaxies. Tuve, it was said, had a shrewd eye for scientific talent. Satisfied with the thirty-six-year-old astronomer's off-the-cuff talk, he handed her a two-by-two-inch photographic plate and asked if she could measure the image's spectra. The spectra came from observations taken by staff member Kent Ford, a young instrument maker at DTM who had just returned from Mount Wilson Observatory, where he had tested an image tube spectrograph, a new tool he'd developed. The tool promised to be a powerful way to capture light streaming toward Earth from distant celestial objects. It could record observations ten times faster than ones taken with conventional photographic plates alone and with the same clarity.

At first glance, the image tube didn't seem that sophisticated. It looked like a series of small film cans glued together, attached to the larger classic spectrograph. But its distinctiveness was in how it managed and collected celestial light. A photon, a particle of light, flying inside it collided with an optically sensitive plate, a photocathode, inside the tube, generating a shower of electrons. These negatively charged particles were focused by magnets and then fell onto a phosphorescent screen creating a glow of light. Photographing that glow instead of the faint light that the distant objects emitted themselves made it easier to analyze the spectrum of the faint stars of a galaxy or the faint photons of any very distant object. In effect, the image tube could make a relatively small-scale telescope as sensitive as the mammoth 5-meter reflector at Palomar.

Eager to have a look at the data the new tool could collect, Vera took the plate with Ford's image tube spectra home and measured it, and when she was done, she immediately dispatched the results to DTM.

It was January 1965.

Vera waited patiently to hear from Tuve. And while she waited, she worked on a paper with the Burbidges, this time trying something a bit different than in past papers. Previously, the Burbidges and Vera typically extrapolated the velocities of stars out past where they could observe as if those stars followed Newton's law of gravity. They assumed the stars' velocities dropped off. But in this paper, Vera, who was heading the research, decided to try a few variations with the data. Her subject was NGC 7331, a spiral much like our nearby neighbor Andromeda, and she was after the galaxy's rotation curve and mass. "I ended up . . . extrapolating [the data] three ways, one with a flat-rotation curve, one with a falling-rotation curve, and one with a rising-rotation curve," she said. "The idea that [the rotation curve] fell was no more secure than the others."[5]

The paper was published in February 1965. The next month, Tuve called and asked Vera how soon she could come by and talk to him about a job. "I can be there in ten minutes," she said.

"No, I meant next week," Tuve said.

"No," Vera responded emphatically. "I'll be there in ten minutes."

She arrived at DTM, straightened her blue suit, and strode confidently into Tuve's office. He was there with another scientist, and they got to talking about the job. Tuve offered a salary. Vera reiterated her terms: she wanted to leave at 3:00

p.m. each day, as she did at Georgetown. Tuve agreed, saying her salary would be exactly two-thirds what she earned at the university.

"I'll take the job," she told him.

Good. She'd start April first, he said.

"I'm teaching at Georgetown."

"Well, I'll call up, and we'll take over your salary and you can still teach," he said. "I want you here on April first."

April Fool's Day could not have been more fitting for Vera to start her new job at DTM. She turned onto the drive (she'd finally learned to drive some years before) that leads to the institute's research campus and wound around its many curves before reaching the parking lot. There, while attempting to back her car into a parking spot, she knocked over some kind of post, either a signpost or a lamppost, which frazzled her. She hurried out of the car and into the administration building to check in. A secretary came to ask Vera for her name and address; not a single piece of paper had crossed the woman's desk telling her of Vera's new position. There was no record of her ever being hired, she was told. Puzzled, both women tried to figure out what was going on. Of course, it was the first time DTM had hired a woman scientist, so that might have led to the confusion. Patiently, Vera struggled through the bureaucratic red tape, then finally found her desk in the office of Kent Ford, the young instrument builder whose spectra she was given at her lunch interview. "She moved in with me," Kent said, and she never moved out.[6] The two teamed up—Vera with galaxies to observe and Kent with his image tube spectrograph—and were ready to head to the telescope.

As Vera slowly adjusted to working at DTM, her demeanor must have changed, even if ever so slightly, but enough for her youngest child, Allan, then only five years old, to notice. One

Figure 9.1
While at Kitt Peak, Vera works with the spectrograph Kent Ford built
for the Department of Terrestrial Magnetism.
Credit: Rubin family.

day he asked her whether she had to pay DTM to let her work
there. He'd pieced together from what was going on around
him that typically you have to pay to get what you want in
this world, and so he wanted to know if his mom had to pay to
work at a place that made her happy. She told him that DTM
in fact paid her to work there. Finally, she felt a step closer to
becoming an established astronomer.[7]

Her contentment may have come from the fact that she
and Ford were preparing for one of their first observing runs. It
was August 1965, and they flew to Flagstaff, Arizona, to spend
a few nights at Lowell Observatory to use their classic 72-inch
Perkins reflector. "Early on, I am sure people thought it was
pretty risqué for the two of us to go out into the night to the
telescope," Kent said, "but we were too focused on making the
image tube work and getting data to think much of it."[8]

Vera recalled that she did have to assert herself just a bit on that first observing run with Kent. "It was his instrument, so he had the right to run the first exposure with it attached to the telescope," she said. As they were setting up for the second exposure, Vera recalled telling him, "Now, it's my turn." She said she felt that wasn't easy to do since he owned the spectrograph. Still, Kent understood. After that, they always alternated taking exposures at the telescope.[9]

Kent's image tube was particularly good at sensing red wavelengths of light, giving the team an unprecedented view of enigmatic astronomical objects called quasi-stellar radio sources, or quasars. These cosmological beasts stole the scientific spotlight after their discovery in the early 1950s because of how much energy they emitted in radio wavelengths. In fact, they were initially called radio stars because they emitted so much radio radiation and had some features of stars. Quasars' optical spectra, finally obtained in the 1960s, revealed the chemical elements of which they were made, but they didn't look like any stellar spectra taken before, so the radio stars were another cosmic mystery.

Slowly, with repeated observations, Dutch American astronomer Maarten Schmidt and others started to piece together a story of what they thought the objects were. Schmidt found the objects' spectra were highly redshifted: the lines were positioned more toward the red end of the electromagnetic spectrum. From Edwin Hubble's famous observations in the 1920s, astronomers knew that the universe is expanding, puffing out like a balloon. As space-time expands, galaxies not tugged on by galactic clusters move farther and farther apart and farther and farther from Earth, stretching the galaxies' light so that when it makes it to our telescopes, it becomes redder. This cosmic expansion appeared to be affecting the light of the quasars, which suggested they were extremely far away. They could be far-off galaxies that outshine any nearby star or galaxy, Schmidt concluded. Not everyone accepted this interpretation of the data, though. Others argued the objects weren't far away at all; but instead were small, bright objects nearby. Time passed, and the debate wore on, sweeping Vera and Kent into the frenzy.

Vera had had an introduction to the quasar debate a little more than a year earlier, in May 1964, when she attended the first Relativistic Astrophysics Symposium in Dallas, Texas. Bob was the one invited to attend—he was studying energy exchange in atoms at the time—but didn't want to go, so Vera went in his place. She was excited to learn more about the quasar work of the time, and its connection to black holes and relativity, but not so thrilled with the overly friendly advances from Yerkes astronomer William Morgan. "It is the only night in my life I've had a problem with an astronomer," she said. He was, in fact, the same astronomer who had rejected the manuscript of her master's thesis, so it was slightly crazy then, when on the night that he was making uncomfortable advances, he asked her why she'd never published the full paper. "I looked at him, unbelieving, because the letter rejecting it is over his signature," she said. Morgan's overly annoying behavior continued, so Vera enlisted her husband's old friend, Ralph Alpher, to not leave her alone with Morgan. Though there was a slight slip-up on Alpher's part (he accidently left her alone with Morgan in the hotel elevator at the end of the night), nothing happened. "That was a very strange evening for me," she said.[10]

Only later would Vera realize that the central topic of the symposium would be a warm-up for her own work on quasars. She and Kent fell headfirst into the quasar research because of Kent's image tube spectrograph. It was very good at detecting wavelengths of light in the red end of the electromagnetic spectrum. It was perfect for studying radio stars. And so together, Vera and Kent started to collect the chemical fingerprints of quasars in optical light. Not long after observing their first quasar, the calls from other astronomers came. Sandage wanted Vera and Kent's spectra on the quasars they'd observed. Others wanted confirmation of objects they'd detected. The pace of research began to rattle Vera, to the point where taking her first trip to Palomar on the long December nights in 1965 gave her a bit of reprieve.

Vera was headed to the California observatory to study the stars of the Milky Way—the other thread of her research at the time—to try to identify the stars' orbital velocities around the galaxy. When Vera arrived at Palomar, she was shown to her room on the second floor of the dormitory where observing astronomers stayed. To get there, she navigated around a red velvet rope cordoning off the stairwell. The rope, she later learned, was there because her room was upstairs, and the other male observers were not to go up to see her.

Vera walked around the rope unfazed. She made her way to her room, unpacked her bags, and prepared for her first evening on the mountain. It was a special night, the night when she would become the first woman legally allowed to observe there. She'd planned to hunt for stars farther from the center of the Milky Way than the sun, following up on the work she'd done with her students at Georgetown. Unfortunately, when she made her way to the telescope, she saw the sky was cloudy. It was cold and snowy that evening, so she wasn't going to get much out of her run with the observatory's 1.5-meter telescope. Caltech astronomer Olin Eggen, who was using the

5-meter telescope that night, wouldn't have any luck either in that weather, so Eggen decided it would be more useful of his time and Vera's to give Vera a tour of the 5-meter telescope (the largest in the world at the time). Eggen slowly walked Vera around the building, showing her the telescope, its controls, and, ultimately, the men's exclusive toilet—the one that had supposedly been the reason women weren't allowed to use the observatory. Eggen, Vera recalled, opened the door with flair, saying, "This is the famous toilet." Of course, the sign on the door said "MEN." That bathroom story and its significance, that women weren't welcome in astronomy, would stick with Vera for decades.[11]

It was an enlightening tour, though the overall observing trip was not as productive as Vera had hoped. She returned from her time at Palomar with no new data on the Milky Way's stars and was a bit stuck in her work on galaxy rotation. She and Kent, however, now found themselves at the center of the frenetic hunt for quasars. The pair worked at Kitt Peak and Lowell with the image tube spectrograph, and in the summer of 1966, they published papers describing the spectra of more than a dozen of the radio objects and confirmed that seven, which Schmidt had studied before, did indeed appear to be as far away as he suggested.[12] Quasars, the data showed, were extremely bright, extremely distant objects, which astronomers didn't quite have the physics to explain, at least not then. What they did know was that if these objects were as far away as they appeared to be and they were extremely active galaxies, then they were a window into the early universe, and quite possibly, they might help to refine the age of the universe once again. With that realization, astronomers launched a cutthroat quest to find the farthest quasar from Earth.

Though many astronomers sought the fame of finding the most distant one, doing it was difficult with standard telescope equipment. It didn't matter that each quasar emits as much light as a trillion suns. The objects are so far away that they were nearly invisible to the telescopes of the time, unless Kent's image tube was attached to them. It was the perfect tool to trap a quasar's faint light, and so the image tubes were in demand, and so was Kent, who traveled great distances to install his tool on telescopes around the world. He'd fly to a telescope in Europe to install the instrument, then fly back and meet his wife at one of New York City's airports. "She'd have a change of clothes for me because I was flying off to Flagstaff to meet Vera to have an observing run," he said. "We started out working on [quasars], and that was a pretty intense thing. We did not miss observing time, for trivial reasons, like installing image tubes."[13]

Because Vera's friends and fellow astronomers knew she had a unique tool to hunt for quasars, they constantly wanted her data. They called her nonstop, asking about what objects she had observed with the image tube and how far away the quasars were. Sandage even wrote Vera a note saying if she found anything new to call him, day or night.[14] It all became a bit too much.

"It was an exciting time, but I was not comfortable with the very rapid pace of the competition. Even very polite phone calls asking me which galaxies I was studying (so as not to overlap) made me uncomfortable," she said. She felt hurried and uncertain about her measurements, especially since she didn't get to the telescope as often as other astronomers to check her work. She feared being rushed would cause her to make mistakes. "That just wasn't the way I wanted to do astronomy," she said. "I would rather drop dead than have another astronomer find I made a mistake."[15]

Feeling her reputation was at stake, she dropped quasar research altogether.

Vera instead turned her attention back to a seemingly less contentious topic—measuring the speeds of stars and gas as they circled galaxies. She set her sights on Andromeda, our nearest spiral neighbor.

10 Andromeda's Young, Hot Stars

On a dark night, far from city lights, you can look up into the sky and see our nearest spiral neighbor. The galaxy shows itself to the naked eye as a faint, grayish oval—a vision of what the Milky Way might look like if seen from a galaxy far, far away. With the aid of a large telescope, Andromeda, or M31, as astronomers call it, morphs into a brilliant poof of gas, extremely bright at the center, then fading a bit into dark and light concentric rings that whirl out from its core. It was these spiral arms that caught Vera's attention.

As she'd spent time with Margaret and Geoffrey Burbidge taking images of distant swirls of stars and dust, Vera had begun to see each galaxy as an individual, with distinctive quirks and beauty, just like each person here on Earth. When Vera first started talking about Andromeda with DTM astronomer Bernie Burke, she wanted to learn more about the way gas moved in the galaxy. In the early 1960s, Burke, along with DTM director Merle Tuve, had started using radio telescopes to study how hydrogen gas moved around the heart of the galaxy. Tuve was fascinated with answering big questions about the cosmos and had a laser focus to collaborate to develop the tools and research programs to answer them; that made DTM a hotbed for astronomy. Burke and Tuve were using data from the 300-foot radio telescope in Green Bank, West Virginia, and the numbers matched well with what Hendrik van de Hulst and colleagues had reported for the movement of gas in the galaxy in the late 1950s using the radio telescopes in the Netherlands. Burke and Tuve wanted to push farther out from the galaxy's center, so they started working with Morton Roberts, an astronomer on staff at the National Radio Astronomy Observatory. In 1966, Roberts published a paper on Andromeda that revealed the velocities of gas clouds that sat as far as 35,000 light-years from the galaxy's center. It was a fraction of the 100,000 light-years that separated the center of the galaxy from its farthest edge. Still, the velocities of the gas didn't appear to drop off as the gas got farther and farther from the center of the galaxy.[1]

It yet was another clue that something with galaxies' motions wasn't quite right, a sign that the spiral had more matter than meets the eye. But, once again, the tipoff didn't seem to catch astronomers' attention. Scientists weren't sure they could trust the radio data. Radio astronomy, after all, was a fairly new scientific discipline, and the researchers couldn't directly observe what was happening with the gas clouds. They continued to cling to the argument that the galaxy's rotation curve would eventually turn over and drop off, just as Newton's law of gravitation suggested it should. Vera, however, refused to accept that assumption.

"I never liked to assume anything," she said.

The radio astronomers' fast-moving gas mirrored the fast-moving stars she'd identified a few years earlier in our own galaxy. Her rotation curve of the Milky Way was flat then. Now, so it seemed, was Andromeda's. "As soon as Burke and then Roberts started to question the idea of declining stellar velocity in

Andromeda," Vera said, "I wondered whether I could turn the telescope at Kitt Peak on our nearest neighboring galaxy and see the same unexpected behavior."[2]

Searching the sky for such unruly galactic action was a welcome diversion from the rush of the quasar work. It was a way to return to a question Vera had wondered about since she'd started studying galaxies more than a decade before. She wanted to know what happens far out, at a galaxy's edges. "Nobody ever seemed to be interested in that," she said. "I really did certainly then feel like I just wanted to work at my own pace, and do something that people would be interested in, but wouldn't bother me while I was doing it."

Kent was on board with the shift. He "was just wonderful, because all he wanted was the image-tube spectrograph to be pushed to its limits."[3] The duo agreed to turn their attention to Andromeda.

For centuries, astronomers have stared at the spiral galaxy, first with the unaided eye, then with ever bigger telescopes, and finally with instruments that could study the galaxy's chemical fingerprints, its spectra. Sir William Huggins took the first spectra of Andromeda in 1890. Then, nearly a decade later, Julius Scheiner used those spectra to argue that Andromeda wasn't a glowing cloud of only gas; it had to be filled with bright, brilliant stars. Decades after that, the galaxy's spectra would reveal even more of its secrets—Andromeda was spinning about its center, just the way the Milky Way does.

To better understand Andromeda's rotation, astronomer Horace Babcock, working as a graduate student at the University of California, Berkeley, used Lick Observatory's 36-inch Crossley reflecting telescope to observe the central regions of our nearest spiral neighbor. Recording the spectra of each region took anywhere from seven to twenty-two hours. But Babcock was patient and persistent. He took spectra after spectra of various regions of the galaxy, an effort that allowed him to finish one of the first large-scale studies of a galaxy's motions. It also allowed him to sketch out Andromeda's rotation curve, which even then, in 1939, didn't seem to drop off.[4]

Following on that work, Walter Baade had used Mount Wilson's 100-inch telescope to observe Andromeda and noticed that it had faint, gaseous patches throughout its swirl. These patches, he concluded, were clouds of gas electrically charged by intense radiation of hot, young stars within them. They were the stellar nurseries of space. Called emission regions, these stellar nurseries were easier to detect in red wavelengths of light because the gas and dust in Andromeda absorbed more of the young stars' blue light—leaving only their red light to travel through space to astronomers' telescopes. Baade started studying Andromeda's emission regions in the 1940s, but the data weren't published until 1964, a few years after his death.

When Vera and Kent Ford turned their attention to Andromeda, they weren't aware of the work Baade had done. They knew they needed to find the galaxy's stellar nurseries, individual cocoons of ionized hydrogen called H II regions, if they were going to draw out Andromeda's rotation curve farther than any astronomer had done before in optical wavelengths, in the light we see with our own eyes. Vera and Kent turned to H II regions because they knew the light of stars far from the galaxy's center would be faint. But the stars' faintness wouldn't be their only drawback. Their spectra were a drawback too. A lone star's spectrum had a lot of dark absorption lines, which covered many wavelengths. Capturing a star's spectra would be tedious and take much more time than taking the spectra of emission regions, which had only a few spectral emission lines that could be quickly computed to the velocities at which the hot, young stars orbit the galaxy.

As a first step, Vera and Kent started plotting out all of the emission regions they would try to observe. Then they went off to Arizona to find them. They started their hunt for the bright, gaseous knots on a bitterly cold night in Flagstaff in 1967. With the lid of the dome at the US Naval Observatory Flagstaff Station open and the 40-inch reflecting telescope pointed to the sky, the ambient temperature in the building dropped to 20 below zero. Vera sat at the telescope trying to identify emission regions, which was difficult, almost like looking into empty space, since the objects were invisible without long telescopic exposures. She'd try for a while, then switch with Kent, who was huddled in an inner room of the building next to a heater warming his hands and feet. She would warm herself for a while, then climb back to control the telescope as Kent went to warm up again. There, as she looked into space, staring up at Andromeda, she saw that "its center had this light, greenish glow," she said. "It always made me wonder if someone, an astronomer in our neighboring galaxy, was looking down on our own and watching us."[5]

Savoring the thought, she switched again with Kent to go warm her hands. Back and forth they went between observing Andromeda and huddling next to the heater. By morning, they'd decided there had to be a better way to find Andromeda's stellar nurseries. "We cleaned up, closed the telescope, and started walking away," Vera said. A bit forlorn, they happened to run into astronomer Gerald Kron, the director of the observatory, as he was coming into his office. Vera and Kent told Kron they were looking for emission regions in Andromeda; it was tedious work, she explained. Then Kron had a thought.

He waved her and Kent into his office and opened a large cabinet. In the drawer were copies of Baade's photographic plates of Andromeda. In sheet after sheet of Baade's plates, there were gaseous knots he'd found many years before. It was a remarkable gift, Vera said, one "truly coming from heaven."[6]

Kron then told Vera and Kent to connect with the photographer at Mount Wilson who could send them photocopies of the photographic plates. In a matter of weeks, Vera had the copies and rediscovered the 688 stellar nurseries Baade had identified when he'd spent hours at the telescope photographing Andromeda. It had taken him twenty hours to capture the light of each individual emission region, but with Kent Ford's image tube spectrograph, it took Vera and Kent less than ninety minutes to record the chemical fingerprints of each emission region. Still, Vera wanted to be meticulous when using Baade's data. When she received the photographs of the glass plates with his data, she spent hours with each one, identifying the exact location of the emission regions in M31.

"For each chosen M31 object, I measured a distance from three nearby stars and made forms for doing the necessary arithmetic at the telescope," she said. The forms gave precise instructions on how to position the telescope with directions: Put star A here, B here, C here. "At the telescopes, with a weak flashlight and freezing hands, we would do the final arithmetic and then move the telescope slowly, so that the invisible object would move to the exact center of our field," Vera said. She and Kent took special care not to confuse an Andromeda star with a faint star in our own galaxy. "Starting from the first try, we never failed to obtain a clear, measurable image," she said. "But it was tedious and required four hands."[7]

Soon Vera and Kent were back in Flagstaff, this time at the Lowell Observatory where the 72-inch Perkins reflector from

Ohio State University was located, and they were intent on exploring Andromeda's stellar nurseries. It was now the late 1960s, and Vera was gone from home much more frequently for observing. Her youngest son, Allan, who would have been a few years shy of ten, started to take notice. "Where's Mom?" he'd ask. "Observing," his sister, brothers, or father would say. They didn't seem particularly concerned that she wasn't there, so he didn't worry. He'd eat the turkey Vera always left for the family when she went to the telescope, and eagerly await her return.

On one particular observing run around that time, Vera and Kent were accompanied by a guest astronomer, who watched intently as they worked. When they were done exposing the photographic plate with the stellar nursery's spectra, they invited the guest to join them as they developed it. Slowly, as if by magic, a fuzzy image appeared on the plate. It was Vera and Kent's first spectrum of a faint emission cloud in Andromeda, and from it, Vera instantly knew that she and Kent would be able to probe the galaxy to depths never explored before. "We were ecstatic but wanted to get back to the telescope to get our next exposure, so our guest volunteered to complete the developing for us," Vera said. "When we returned with the second plate, we found that he had accidentally washed the first plate in hot water rather than cold." The emulsion was destroyed, the data gone. "The glass was absolutely blank," she said. "Our guest was mortified, but I was so delighted that everything worked that I didn't care. I knew we would need hundreds of exposures."[8]

Losing one exposure was of little consequence.

Vera was right not to wring her hands at the lost plate. After the observing run at Lowell, she and Kent loaded up their equipment in a white Suburban provided by the observatory and headed south on a scenic five-hour drive with no interstates. They were headed to Tucson, to Kitt Peak National Observatory. Along the way, they passed acres of Japanese flower farms, and Vera asked to stop. She bought a 25-cent bouquet of sweet peas, she recalled, an early hint of her love of gardening.[9] Flowers in hand, Vera hopped back in the car, and she and Kent headed west on the Tucson-Ajo Highway for some fifty miles and arrived at the observatory. There, a group of Native American employees were waiting, prepared to help unload Vera and Kent's gear. "There must be a better way to make a living," Vera remembered them saying as they stared at the truck. "I never knew if they were referring to themselves or to us."

Without another word, the team unpacked the image tube and got the pair ready for another night of observing, this time with Kitt Peak's 84-inch reflector, the largest on the mountain. The first spectrum came in, then another and another. Vera was developing the plates as soon as they were exposed. She'd eat an ice cream cone while waiting for each one. Once she'd developed the photographic plate with the spectral data showing emission lines at specific wavelengths of light from the gas surrounding the hot, young stars, she'd measure the location of a specific line called hydrogen-alpha. The line forms when the electron in a hydrogen atom loses energy; specifically it drops from the third to the second lowest energy level of the atom. Vera would measure the location of the line to the nearest thousandth of a millimeter. Then she'd compare that position to a spectrum exposed on a plate in the laboratory before it was exposed to starlight at the telescope. Determining the difference in the position of the lines, she could calculate the velocity at which each stellar nursery circled Andromeda's center.

Four ice cream cones later, Vera had enough velocities of emission regions to plot them against the gaseous knots' distances from the galaxy's center. She and Kent drew out Andromeda's rotation curve.

It was flat.

Vera stared at it in amazement. She was shocked to see it, yet she could recall the voices of criticism she'd heard earlier regarding her earlier work on galaxies. To confirm what they saw, Vera and Kent collected more data. By 1968, they'd gathered spectra of sixty-seven emission regions in Andromeda, the closest about 10,000 light-years from the galaxy's center, the farthest 79,000 light-years from the center. Yet again, plotting the speeds of the gaseous knots against their distances from the galaxy's center, the curve rose steadily, then leveled off right around 200 kilometers per second, roughly 45,000 miles per hour. It did not drop off.

Andromeda's rotation curve was flat.

In December 1968, Vera presented these results at the AAS meeting in Austin, Texas. After her talk, legendary Caltech astronomer Rudolph Minkowski approached her and asked her questions about the results. He then asked Vera when she would publish a paper on the data.

"I don't know," Vera responded. "There are more regions we could observe."

"I think you should publish it now," he said sternly.

Heeding the advice after returning from Austin, Vera spread her data on the long, wooden dining room table where she, Bob, and the kids always worked on projects and schoolwork and set to writing up the results. Watching her work and seeing that she took such joy in it, Allan asked what she was doing. She told him she was studying a galaxy called Andromeda and that it didn't seem to behave as astronomers thought it should. She "would always talk about her work during dinner," Allan said. "It was very accessible."[10]

A little more than a year later, in February 1970, Vera and Kent reported Andromeda's flat rotation curve in the *Astrophysical Journal*.[11] There was no mention of any extra matter or an explanation of the unexpected speeds; the paper just presented the data showing that the speeds of the stellar nurseries did not drop off as they were supposed to, according to Newton's law of gravitation.

A few months later, K. C. Freeman, a well-connected astronomer in Australia, published a paper in the same journal. He also found flat rotation curves for two other galaxies, and as a result, he argued that galaxies have two parts: a main sphere where most of the stars and gas sit and a disk that extends farther out that is much darker. What the disk was made of, Freeman said, wasn't clear, but he suggested that it might account for galaxies' rotation curves not falling off as expected.[12]

In the appendix of that work, Freeman elaborated further, a discussion that arose from talks with radio astronomer Mort Roberts. Freeman noted that if the velocity data were correct, "then there must be in these galaxies additional matter which is undetected." The mass, he continued, "must be at least as large as the mass of the detected galaxy, and its distribution must be quite different than what's seen in the optical wavelengths of the galaxy."[13]

Freeman's paper, revised in December 1969, before Vera published her data, failed to cite her work. But the fact that the papers were published in such quick succession proved Minkowski's point: the field was competitive, more competitive than Vera ever expected, which could be why Minkowski urged Vera along.

As the papers with flat rotation curves rolled out, Vera and Kent also began collaborating with Mort Roberts. The three of them had begun measuring the velocities of stars and gas in galaxies other than Andromeda, drawing each one's rotation curve, and comparing them with the curves Roberts had drawn from the radio data. The curves all appeared to be flat.[14]

This was a fact that came up at family dinners when Bob and Vera would discuss her work. Even after the meal ended, they'd continue to talk about their research, Allan Rubin recalled. "I understood that after dinner they could do whatever they wanted, and what they chose was to sit at the large dining-room table (used for dining only when we had company) and continue to work on the same science that they had been working on during the day," he said. "It was pretty clear that what they were doing for a living was what they most wanted to do."

Bob and Vera's attitude toward science was catching. By that time, the early 1970s, the Rubins' eldest son, David, was starting to develop a serious interest in geology. And after Vera volunteered her time at her kids' high school, Judy, their daughter, took a liking to astronomy. (She was a Westinghouse National Talent Search winner in 1970.) Not to be left behind, the Rubins' two younger boys, Karl and Allan, were showing a knack for mathematics.

Vera's excitement to discuss her work didn't stop with her family, though. She spent hours chatting with Kent Ford, Mort Roberts, and another astronomer, Sandra Faber. A few years earlier, Faber (Sandra Moore then) had spent the summer working with Vera and Kent at DTM when she was an undergraduate at Swarthmore College. That summer, she'd built instruments, measured spectra, and published a paper on the Virgo cluster. Then she'd gone off to Harvard to start her graduate course work, and once she'd finished, she found herself in a peculiar situation.

She'd gotten married and her husband, Andrew, was in a master's program studying applied physics at Harvard. He'd been given a deferment from the Vietnam War to start his degree, but the deferment was only for a year, so he and Sandra had to choose: move to Canada, send Andrew off to war, or have him find a job that would give him another deferment.

Having studied underwater acoustics, Andrew landed a position at the Naval Research Laboratory in Washington, DC. It was the best of the outcomes, but there was a hitch: Sandra was still in school at Harvard and didn't want to be in a long-distance marriage. She discussed this with her adviser, and they agreed that since she was done with her class requirements, she could move to DC and work on her thesis there. She initially arranged to set up an office with the astronomers at the Naval Research Lab and got to work studying galaxies. Pretty quickly she realized, though, that there was really no one at NRL she felt she could talk with about her work. She found the atmosphere isolating. Hearing this, Vera and Kent invited Faber to work with them again.

"Overnight I had found a very pleasant and productive niche," Faber said.[15]

One day while working, she recalled, the phone rang. Vera answered. It was Mort Roberts.

"Vera," he said, "I have some very strange results on rotation velocities in M31, and I'd like to come up and discuss them with you." The next week, Roberts was at Vera and Kent's DTM office sharing the results.

The four—Faber, Rubin, Ford, and Roberts—sat around a table and looked at the latest radio data on Andromeda. Roberts had drawn a new rotation curve for the galaxy that extended thousands of light-years farther than what he'd measured before, and the velocities of the gas clouds way out from the galaxy's center were still the same as the ones farther in. His rotation curve was still flat.

Roberts's excitement was evident as he explained his latest observations.

"Well, so what?" Faber said. "You know. Fifty percent farther out is only 50 percent more mass. It's not a big change in mass."

"You don't understand," Roberts replied. "There's no light there."

Faber still wasn't impressed. "I was already convinced that there was something totally crazy about mass measurements in astronomy," she said, and she was also convinced that "somehow these velocities didn't mean anything, that they were wrong."

She wasn't totally off base either, she thought. She'd read a lot about the structure of galaxies and the nature of groups of galaxies to find her thesis project, and she had planned to study double galaxies, looking for whether the galaxy duos had more shared features than two galaxies chosen at random. She'd combed the galaxy catalogs and looked at double galaxies that had huge differences in velocities. And when she'd used the velocities to calculate the galaxies' masses, she'd found they were huge; they couldn't possibly exist based on what astronomers knew about physics. "They were impossible," she said.[16]

She had, of course, read about Zwicky, Smith, and others finding what could be excess mass in clusters of galaxies, and she attributed the impossible masses in the double galaxies to the same phenomenon, whatever it was. Still, she just didn't think it was worth dealing with.

"Nobody had ever solved this problem. And I just didn't want to bother with it. It didn't seem to be a question that was ripe for solution," she said. So when Roberts came and showed the group "yet one more crazy velocity," she'd thought, *Why are you making such a big deal? It's just one more example of this problem.*

Faber's skepticism wasn't the exception in the astronomy community. It was the rule—at least, until Vera and others found flat rotation curves in more and more galaxies.

11 The Matter We Cannot See

Sandra Faber wasn't alone in casting aside flat rotation curves as an unsolvable anomaly. Lots of other astronomers didn't think the data were right either. And although the evidence had been building for years on many different fronts, most researchers still felt that the universe wasn't missing any mass. To be fair, these astronomers were going on precedent. It's hard to convince people something exists when they can't see it.

Such doubt is a persistent problem in science. Consider the case of black holes. Predicted to exist as early as 1784, then expounded on again in the early twentieth century, the insatiable eaters of outer space seemed "like unicorns and gargoyles . . . more at home in the realms of science fiction and ancient myth than in the real Universe," Caltech theorist Kip Thorne once wrote.[1]

Black holes were fantastical beasts for which there was no solid evidence until 1971, when scientists spotted X-rays coming from a blue star whipping around a strangely dark spot in space. The X-rays, the scientists speculated, were coming from matter being torn from the blue star by the dark object, something that had the features of a black hole. It was the first accepted evidence of black holes' existence. The discovery was made right around the time Vera and others were scratching their heads over the flat rotation curves. And just as astronomers had undercut the existence of black holes without solid evidence, they debated the reality of flat rotation curves and the existence of large quantities of matter we could not see.

Despite such doubt, scientists like Vera and Mort Roberts dove head-first into searching for additional evidence of flat rotation curves. Seth Shostak, then a Caltech grad student, also decided to take a crack at finding evidence for and explaining flat rotation curves. As an undergraduate at Princeton, Shostak had worked with physicist David Wilkinson and was inspired by Wilkinson's enthusiasm for detecting the relic radiation of the big bang. Shostak carried that inspiration with him when he moved to Caltech and decided to study galactic rotation curves after talking with one of his advisers, Maarten Schmidt—the discoverer of the optical signatures of quasars. Intrigued with the idea of collecting radio data to draw out galaxy rotation curves, Shostak teamed with Caltech radio astronomer David Rogstad and combined the data of two linked 90-foot radio dishes in Owens Valley, California. They combined the dishes to form an interferometer to measure the rotational velocities of galaxies using the galaxies' 21-centimeter hydrogen line emission feature. Shostak and Rogstad measured the amount of redshift on both sides of the galaxies and were able to use the data to gauge the velocities of gas there and how the velocities of gas changed with distance from the center of each galaxy.

Shostak and Rogstad had been following Mort Roberts's work and by using the 90-meter radio dishes as an interferometer, they gained impressive spatial resolution that allowed them to zoom in on gobs of gas at a galaxy's farthest edges and extend a galaxy's rotation curve out to the point where there's little luminous matter left to give off light. With the data, Shostak drew the rotation curve of three galaxies, the most prominent being NGC 2403, a spiral in the constellation Camelopardalis. He found that the gas at the farthest edges of the galaxy swung around it just as fast at the gas closer in. That galaxy's rotation curve didn't drop off.

Like Faber and many others, he too initially thought it was a very curious result, one that demanded an explanation. Shostak, who'd read Vera's paper on Andromeda and Freeman's on two other galaxies, took a shot at giving one: the only way the rotation curve could be flat that far from a galaxy's core is if there were "underluminous material in the outer regions of the galaxy," he wrote in his PhD thesis, completed in 1971.[2] His idea of underluminous material seemed to align with what Freeman had proposed a year earlier: that galaxies with flat rotation curves had to have extra mass, matter that we couldn't see.

During Shostak's long nights running the radio telescopes to collect his thesis data, he said he realized the big dishes peering into the night sky were the perfect tool for another project: searching for signs of extraterrestrial life. So intrigued with the idea, he even dedicated his thesis to NGC 2403 and all of its inhabitants.

After finishing at Caltech, Shostak moved to the National Radio Astronomy Observatory (NRAO) in Charlottesville, Virginia, where Roberts was on staff. There, Shostak continued his work with Rogstad on galaxy rotation curves, and in March 1972, they published a paper on the features of five spiral galaxies, including the velocities of neutral hydrogen in their gaseous arms. The galaxies' rotation curves were all flat.[3]

Shostak, however, wasn't satisfied. He had more data to analyze, and so that summer, he hired an intern, an astronomy major at Radcliffe College, to help. The intern was Judy Rubin, Vera and Bob's daughter. Judy most likely discussed Shostak's research with her mother, especially considering how much Vera enjoyed talking about astronomy. "I'd find it kind of hard to believe Judy wasn't talking to her mom about what she was doing all summer," Shostak said.[4]

By this time, word was getting around that galaxy after galaxy appeared to have flat rotation curves and that those curves might be explained by mysterious, or at least subluminal, matter. As Shostak continued his radio observations, Vera and Kent were finishing their in-depth analysis of gas and stars in Andromeda, and Vera decided to return to the project she'd done with her students at Georgetown in the early 1960s, where she tried to identify faint, blue stars far from the center of the Milky Way. She could use those stars to draw out our own galaxy's rotation curve, she reasoned, and now she didn't need to rely on others' data. She could collect everything herself. To advance the work, she made measurements of Milky Way stars' speeds using the 1.2-meter telescope at Palomar Observatory, an extension of the project Allan Sandage had asked her to do there nearly a decade earlier. Back then, Vera had identified eighty-nine stars to follow up on. It would be those stars that she could target, take their chemical fingerprints, and ultimately draw out the Milky Way's rotation curve. This curve had been neglected since Vera's last study in 1965, and she decided it was time to take another look at the movements of stars in our own galaxy.[5]

As Vera homed in on the Milky Way's stars, Mort Roberts worked on getting rotation curves of more galaxies far beyond our own. He'd been invited to the Kapteyn Astronomical Institute at the University of Groningen, and there he used Oort's Westerbork Observatory, with its 25-meter radio telescope, to study a fuzzy spiral called M81. With Arnold Rots, Roberts measured the speeds at which gaseous clumps of neutral

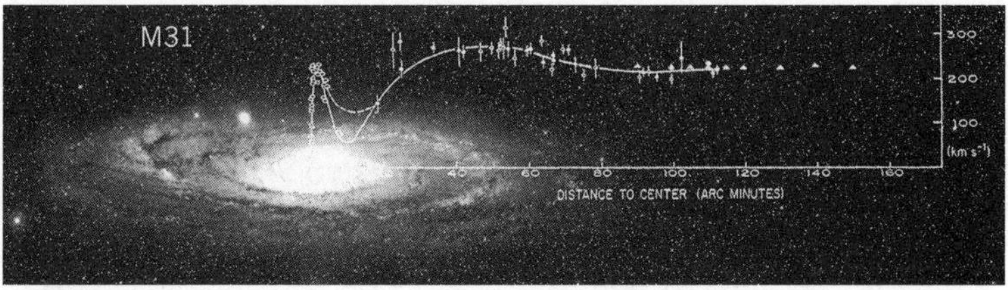

Figure 11.01

An image of Andromeda overlaid with the galaxy's rotation curve—a plot of the velocities of stars and gas orbiting the galaxy against their distance from the galaxy's center. Radio astronomy observations are shown as triangles. The curve is flat even far beyond where the starlight drops off, implying that there must be extra mass—dark matter—in the outer reaches of the galaxy to pull the stars and gas along at faster than expected speeds. Courtesy Carnegie Institution for Science/Vera Rubin and Janice Dunlap.

hydrogen circled the galaxy's center. He also pulled in data on the galaxy taken at NRAO. Comparing the rotation curve of M81 with the curve of Andromeda from Vera and Kent, and M101 from Shostak and Rogstad, Roberts and Rots could take a detailed look at how stars and gas move in different types of spirals.

Based on the data, Roberts and Rots concluded that the spirals seemed to harbor similar amounts of matter—roughly 200 billion times the mass of the sun. But they noted that the galaxies have slightly different shapes. Andromeda, for example, has a bright center and a few spiral arms rippling outward from its core; M101 has a more consolidated core compared with Andromeda and more spiral arms; and M81 has a fuzzier, seemingly larger, diffuse core than Andromeda and fewer spiral arms.

Because the galaxies are swirls of stars with slightly different structure, an analysis of their rotation curves might reveal whether it was common for gas and stars in spirals to move around galaxies much more quickly than expected. Flat rotation curves, critics had said, were unique to Andromeda, so this study would show if that was, or was not, the case. Subtle differences between the rotation curves did exist, Roberts and Rots noted, depending on the galaxy's spiral shape: M81, the galaxy with the bulkier core, had a rotation curve that rose rapidly, then tapered off. M101, with more spiral arms, had a curve that rose steadily, then turned over and flattened. Andromeda's curve had a big rise, then dipped a little and leveled off. Nevertheless, in every case, the researchers wrote, the rotation curves "decline slowly, if at all."[6]

Roberts and Rots's paper, along with Vera and Kent Ford's on Andromeda and Shostak and Rogstad's, caught theoretical astrophysicists' attention. They were captivated by the flat rotation curves and also curious about what the speeds of stars and gas meant for the evolution of galaxies' spiral arms. In these arms, streams of interstellar gas, both neutral and electrically charged, slam together, triggering star formation, as waves of dust wander through. In the early 1960s, Alar Toomre of MIT and colleagues questioned the way spiral arms formed and then became stable as the stars and gas swung around the galaxy's core. Toomre ultimately concluded that the stars probably streamed around a galaxy's core in a smooth, uniform, circular fashion. If any stars or gas deviated from that circular motion, the disk, the flat pancake holding the stars and gas, would start to break apart. Using this set of assumptions, he was able to calculate just how much the velocities of the stars could deviate from their standard swirling speed to suppress any instability in the disk and prevent it from falling apart. What he found was that the speeds of stars in our

Plate 1

UGC 2885 may be one of the largest galaxies in the universe. Near the end of her career, Vera Rubin was still studying this behemoth spiral to search for signs of dark matter's existence at the very edges of the galaxy. Because of her work on it, the spiral is called Rubin's galaxy. Credit: NASA, ESA, and B. Holwerda (University of Louisville).

Plate 2

The Bullet Cluster is one of the more recent pieces of evidence of the existence of dark matter. Credit: X-ray: NASA/CXC/CfA/M. Markevitch et al.; Optical: NASA/STScI; Magellan/U. Arizona/D. Clowe et al.; Lensing Map: NASA/STScI; ESO WFI; Magellan/U. Arizona/D. Clowe et al.

Plate 3
In the decades to come, astronomers will scan the night sky using the
Vera C. Rubin Observatory in Chile. Credit: Rubin Obs/NSF/AURA.

Plate 4

Vera sits with her children (left to right) Karl, Dave, Allan, and Judy at Bear Lake in Colorado's Rocky Mountain National Park in 1961. Credit: Rubin family.

galaxy, near the sun, didn't deviate too far from the standard orbital speed, which kept our galaxy's disk stable.[7]

The analysis was based on relatively small samples in the galaxy, however, and it wasn't clear it would hold up as astronomers started to measure the velocities of the stars farther and farther from the galaxy's center. What was also unclear, as Toomre had noted, was accounting for other factors that might affect the velocities of stars, such as the stars' gravitational interactions with nearby clouds of gas and dust. Added to that were differences in the gravitational tug of galaxies' central bulges; some were brighter than others, suggesting they had more mass, which might affect how a galaxy's stars and gas moved around its midsection.

In the late 1960s, when Vera started working on Andromeda's rotation curve, electronic computers were slowly becoming sophisticated enough to simulate the motions of stars in the disk of a galaxy; the computer would represent the stars as particles in a disk-shaped blob. At NASA's Langley Research Center in Hampton, Virginia, Frank Hohl had started using the facilities' electronic computers to calculate what would happen to large numbers of particles over time, a simulation of stars in a galactic disk. In a 1970 technical report, he described how running these simulations, integrating the equations of motions for thousands of particles, on a Control Data 6600 computer system countered astrophysicists' notion of stability in the disk. Their earlier idea of the way the disk remained stable didn't work. In fact, the model "was found to be violently unstable," Hohl wrote in his report.[8] Kevin Prendergast, who had worked with Margaret and Geoffrey Burbidge, and on occasion with Vera, found something similar. In his simulations, the disk the particles were in disintegrated over time.

His team even tried tweaking the particles so they'd represent cooled gas. That didn't help; the disk of stars still didn't last in the simulations.[9]

Ultimately Hohl tried something in his simulations that did finally gave him a spiral-like structure in his disk of particles. He added what he called a fixed, central potential, a cloak of matter encompassing the disk of particles. He told the computer to run the simulation of the particles over time and watched what happened. The particles buzzed around the disk and swirled out into spiral-like arms. And this time the simulation revealed that for that spiral structure to take shape, 10 percent of the mass of the system had to be in the stars and the rest had to be in the cloak of mass wrapped around them.

In running the simulation again and again, the particles jittered, then ultimately settled out into a pinwheel shape. The simulated galaxies looked very similar to the swirls of spiral galaxies seen in space. What's more, this structure evolved with the particles having motions around the center of the system similar to what was observed in actual galaxies.[10] It was a tipoff that a "massive subluminal halo might account for the stability of the disk of stars in a spiral galaxy," recalled theoretical astrophysicist Jim Peebles.[11]

As a graduate student at Princeton in the 1960s, Peebles had become interested in the early universe and was part of the team that confirmed Arno Penzias and Robert Wilson's discovery of the cosmic microwave background, the relic radiation of the big bang.[12] There had been a hint that this relic radiation existed as far back as 1941, a discovery made by Canadian astronomer Andrew McKellar, but it was Penzias and Wilson's discovery that offered convincing evidence that there was indeed a big bang at the beginning of cosmic time and that we are living inside its remnant radiation field. The discovery was revolutionary not only for providing evidence for figuring out what happened in the instant the universe formed; it would also become essential to understanding how matter was distributed in galaxies and throughout the universe itself. Working on confirming the discovery, Peebles began thinking about what ingredients must have existed in the early universe for clumps of gas to form to create stars and then for those globs of gas to coalesce into galaxies and clusters of galaxies. This was especially important to consider since the universe was expanding. Matter would have been pushed farther and farther apart, so how could it still condense enough to form the vast galaxies seen today?

After earning his PhD, Peebles had joined the physics department at Princeton as an assistant professor and after a few years had taken a sabbatical at Caltech. This was about the time that evidence for flat rotation curves was growing. While Peebles was in California, he began to wonder if he could use computer simulations to trace the evolution of galaxies from their early state of gaseous globs through their metamorphosis over cosmic time into the clusters seen today. Peebles was so curious about this galactic metamorphosis that, on his drive from Caltech back to Princeton, he stopped at Los Alamos National Laboratory, a center for nuclear weapons research, where he was invited to use the lab's CDC 3600 computer.

Peebles planned to use the computer to integrate the equations of motion for hundreds of particles; the particles would represent pockets of gas in the early universe—what astronomers call protogalaxies. The computer simulations would reveal whether the notion Peebles had for early galaxies and their interactions would align with what astronomers saw in the actual cosmos. Under the watchful eye of a lab employee, Peebles, who was Canadian and not to be left alone at the lab,[13] let the simulation play out. It showed that protogalaxies could be moving away from each other early on, then eventually fell back together, pulled closer by some strong gravitational force. In fact, the simulations, given relatively vague initial conditions of the early universe, churned out a cluster that looked strikingly similar to the Coma cluster, a hint that Fritz Zwicky's description of the cluster way back in the 1930s might have been right. The simulations seemed to suggest there could be some unseen stuff keeping those galaxies in that cluster stuck together.[14]

Intrigued with the computational method, Peebles brought it back to Princeton and linked that work with earlier work he'd done on what caused galaxies to spin around a central point. He'd previously shown that as a protogalaxy forms, the matter within it interacts with the mass outside it, causing the protogalaxy to rotate around the mass at its center. So maybe the early seeds of galaxies spun. Would that be what kept them spinning billions of years later? Jan Oort didn't think so. He'd asked that question and found that that circular motion early on wouldn't be great enough to keep a galaxy whirling around itself for billions of years. So there was a discrepancy there. And there was another problem with Peebles's simulations: his galaxies, like Hohl's earliest ones, kept falling apart. The mass in the disk of a galaxy might not be enough to keep the galaxy together, Peebles concluded. Eager to resolve the issue, he started talking about it with another Princeton astrophysicist, Jeremiah Ostriker.

Ostriker had been investigating the rotation of dead stars called white dwarfs, and he too had hit a frustrating dead-end: in his calculations, the stars would spin into a pancake-flat disk and then disintegrate. But real white dwarfs, the ones out there in space, obviously weren't falling apart completely as they rotated. Something was missing. As they talked, Ostriker and Peebles realized they were facing the same problem: if they put mass, whether from a dead star or a galaxy, into a spin, the star or galaxy would fly apart. Peebles, however, had had success with the simulations he'd done at Los Alamos, ones that hinted that some extra mass might stabilize galaxies. Peebles and Ostriker then ran simulations again and enveloped the particles representing galaxies in spherical halos. Similar to Hohl's central potential, this spherical halo also took the form of a cloak of mass enshrouding each simulated galaxy. The galaxies became much more stable; as they made one, two, and three rotations around their center of mass, they didn't fly apart as the galactic particles without the halo did. What that suggested, Ostriker and Peebles wrote, is that a galaxy might have an added component: a bright central bulge; a disk of gas, dust, and stars; and a halo that envelops the galaxy. The halo would have the same mass as the disk, or it would have up to 2.5 times the mass of the disk. Either way it's that halo, they suggested, that might explain the flat rotation curves that Vera, Roberts, and Shostak were finding.[15]

Connecting the halo mass with the mass to explain rotation curves was a "major step," Peebles said, and it also accounted for motions of galaxies in groups and clusters.[16] It was a solution to the problem that Faber thought was unsolvable, and it was a sweeping reconstruction of galactic architecture. Now, not only did galaxies have a bright, central core and a pancake-shaped disk, but also a spherical halo of matter. The addition of the halo was even more of a concrete argument that galaxies and the universe had more matter than we could detect at the time.

By then, Ostriker and Peebles weren't alone in their thinking. As they worked on their idea of a subluminal mass lurking about each galaxy, a group of researchers led by Jaan Einasto at the University of Tartu in Estonia was also thinking about what it would take to explain the odd behaviors of galaxy clusters, specifically the problem Zwicky had pointed out forty years before. Zwicky had said there had to be more mass than astronomers could see in the Coma cluster to keep its galaxies together; Einasto and colleagues now agreed, arguing each galaxy had to have some unseen element, what they called a corona, to keep everything together, and that meant, without reservation, that "clusters of galaxies must contain dark matter," they wrote. That dark matter (the first use of the phrase in the modern era) might take the form of "massive and extended coronas of unknown origin." The coronas, it turned out, were similar to Peebles and Ostriker's galaxy-encompassing halos.

To deduce that galaxies needed these unseen coronas, Einasto and his team also ran computer simulations, using data on 105 galaxies and modeling them as test particles. The team's goal was to estimate the mass and radius of a galaxy's halo compared with the mass and radius of its stellar populations, and the simulations showed the mass and radius of the corona exceed the mass and radius of known stellar populations of a galaxy by one order of magnitude, the researchers reported in 1974 in the journal *Nature*.[17] The coronas were big, they said, much bigger than the disk of stars itself, and the coronas added mass to the galaxy. That added mass reduced, or nearly wiped away, the mass discrepancy Zwicky had noted in his calculations of the Coma cluster. The coronas, or halos, seemed to solve Zwicky's problem.

Just a few months later, Ostriker and Peebles, with Amos Yahil, again added to the discussion, this time tying the computer simulations on galaxy interactions to the need for more mass in groups and clusters of galaxies, the flat rotation curves of spiral galaxies, the stability of the disks of spiral galaxies, and features that astronomers had observed in merging spiral galaxies. As a result, the authors of both papers came to the same conclusion: there had to be more matter in a galaxy—about ten times more matter—than what we can see from its starlight and that the matter had to increase farther and farther from the galaxy's center.

The idea that a galaxy needed more matter than we could see was finally taking hold.

Of course, "It was not new news," Ostriker said. "We had just put it all together."[18]

The conclusion was exciting, he explained, because it allowed astronomers to recalculate the density of matter in galaxies. That calculation was important for understanding the fate of the universe by allowing astronomers to estimate a critical value they called omega. Omega offers a glimpse at the overall shape of the universe and what will happen to it billions and billions of years from now. Estimating omega using only the mass of galaxies based on visible starlight, astronomers came up with a value of 0.02. Factoring in the matter we cannot see, described in the theoretical work of Peebles, Ostriker, and Yahil and Einasto and his colleagues, omega jumped to 0.2. That addition might seem insignificant, but it meant that the density of matter in the universe was 20 percent of the critical cosmological density, not just 2 percent of that density. At 100 percent, or omega equals 1, the universe is flat, like a feather quilt stretched taut. Its expansion slows, approaching zero but never quite stopping. If omega is above 1, the universe is spherical. It will expand and then contract, falling in on itself in a big crunch. With omega less than 1, the universe is shaped like a horse saddle. It would expand forever, ultimately becoming cool and dark as stars slowly lose all of their energy and flicker out.

This cold close of the cosmic curtain appeared to be the universe's fate if subluminal matter didn't exist. And if invisible matter did exist, and it existed in the quantity the theorists calculated, then there was still a solid chance that the cosmos would slowly fade to darkness, but there was far less certainty of that final fate.

Even with all the calculations and simulations, resketching the structure of galaxies to include an invisible halo was a revolutionary shift in thinking. It was a bold reimagination of the universe, an extraordinary claim that would require extraordinary evidence, as Carl Sagan would later say.[19] And so as Peebles, Ostriker, and Yahil wrote in their 1974 paper, hunting for "giant halos surrounding ordinary galaxies . . . would be especially rewarding."[20]

Who would hunt for hints of those haloes? Vera, of course. Nearly a decade earlier, she'd wanted to know what happened to the speeds of stars at the outer edges of our galaxy, and then she'd again asked the same question for Andromeda. Very few people were interested in addressing that question, which she had liked, and with this theoretical work, she had permission to pursue it further, to truly see if the phenomenon of flat rotation curves held in other galaxies and if they too might have dark matter.

12   More Matter Than Meets the Eye

On a cold December day in 1980, a postcard addressed to Vera arrived at the Department of Terrestrial Magnetism in Washington, DC. On the front of the card was a drawing of a vast, ornate dining hall. Rows of tables and benches lined the room. Chandeliers dangled from the ceiling, and portraits of famous men hung on the walls. The scene was from Corpus Christi College at Cambridge University in England. On the flip side of the postcard was a note from Norbert Thonnard, Vera's collaborator. Near the bottom was a striking line: "Theorists have finally accepted flat rotation curves!"[1]

Norbert was in Cambridge, attending a meeting at the university's Institute for Astronomy. He told Vera the days had been packed with lively discussions after astronomers' presentations of new results. He had presented work on galaxies, specifically confronting the audience with the idea that stars farther from a galaxy's center whip around it just as fast as stars closer in.

In the past decade or so, when she, Norbert, or others discussed the stars and gas and explained that the speeds at which they orbited a galaxy's center didn't change as they got farther and farther from it, other astronomers begged to differ.

There'd been some uncertainty in the measurements. Some radio astronomy data even suggested that gas farther out in some galaxies didn't move around their cores as quickly as gas closer in. Curious as to why there were such discrepancies, Vera, Kent, and Norbert had begun tracking the speeds of stars and hot gas in one galaxy, then another, then another. In the dozen or so they'd studied in the 1970s, stars far out and those far in traveled around their galaxy's core at similar speeds. And now, finally, in 1980, others were beginning to believe the data. The postcard was proof.

In the decade before Norbert sent Vera the Cambridge postcard, she and others with whom they collaborated spent countless hours at their telescopes collecting data on galaxies' hot gas and stars, searching for evidence of flat rotation curves and, consequentially, subluminal matter. Even after 1974, when theorists had begun to agree that galaxies had subluminal halos, Vera still looked for more evidence, this time starting with galaxies like the Milky Way. Her intention was to determine if she could glean anything from those spirals that could be applied to understanding the rotation of our own. At the time, University of Texas astronomer W. L. Peters III, a colleague of Vera's former collaborator Gérard de Vaucouleurs, provided data that implied the Milky Way's spiral arms met not at a spherical core but at a bar at its center.[2] Again, being inside our galaxy makes it hard to see what the Milky Way looks like from the outside, but Peters's analysis suggested that if we could take a spaceship and fly outside our galaxy, then look back at it, we'd see a spiral with a line of stars streaming through the galaxy's center. NGC 3351, a galaxy that sits 32 million light-years from Earth in the constellation Leo, looks a lot like this. Since it had a similar structure to what the Milky Way might look like, Vera decided to study its stars, along with the hot gas in the galaxy's stellar nurseries. Tracking the speeds of the stars in the bar of NGC 3351 revealed they moved about the galaxy rather slowly, which suggested the bar was a "quasi-stationary feature of the galaxy," while the material farther out moved more quickly.

The barred spiral's rotation curve was flat.[3]

The rotation curves of barred spirals NGC 5383 and NGC 5728 were also flat, Vera found. It was the clue that flat rotation curves were probably common among galaxies, but Vera wanted more data to be sure. She kept looking at galaxies.

As deeply interested as she became measuring spirals' rotation curves, her research was never single-threaded. She regularly launched new projects based on her curiosities. One she did with Bob. Together, they scoured documents for the first historical references of the Crab Nebula, which was seen as a stunning "new star" or nova in 1054 A.D. that lasted for several years until it faded away. Resembling a gaseous fireball today, or an explosion of fireworks, the nebula sits 6,523 light-years from Earth in the direction of the constellation Taurus. After hours reading related literature, the Rubins's detective work revealed, without a doubt, that John Bevis was first to describe the cosmic lightshow in 1731. Others had suggested an earlier reference, but when the Rubins reviewed the archival material making the claim, they found there had been some confusion between the constellation Cancer, "the crab," with the Crab Nebula.[4]

"Everybody found it very amusing," Vera recalled. "It was a cute little story." The paper describing the historical references was the only one the Rubins would write together as coauthors, though Bob often ended up in the acknowledgments of many of Vera's other papers. The same year that Vera and Bob presented their work on the Crab Nebula, she, along with Kent

Ford and the Rubins' daughter, Judy, published a paper revisiting the questions Vera had asked in her master's and PhD theses—whether galaxies have motions aside from their movement as a result of the Hubble constant and if galaxies ever congealed into supersized clumps. There was lots of chatter in the literature about how big galaxy clusters could be, but the team decided to gather data on galaxy groups, specifically the way galaxies move relative to each other. They were after the clusters' velocities. And their work here would ultimately bring Vera back to rotation curves, possibly because the topic, again, was contentious.

When the trio looked at their data on galaxies as they moved through two different regions of space, they found the groups moving at very different velocities than expected. The anomalous data offered a hint that these galaxies might be clumped together in larger clusters than astronomers expected. Or one group might be moving in some coordinated way toward the other, motion that was vastly different from what was predicted from data on the relic radiation of the big bang. The data even suggested that the Hubble constant might not be constant across the sky and could be 1.25 times faster in one group of galaxies compared with the other.[5]

Astronomers weren't buying the data or the team's interpretation of it. In a paper published in the winter of 1976, astrophysicists Laurent Nottale and Hiroshi Karoj called the data anomaly the "Rubin-Ford effect" and argued it might be explained away by changes to the light emitted by the galaxies studied.[6]

A few months later, Cambridge astronomers S. Michael Fall and Bernard J. T. Jones also took aim at the Rubin-Ford effect, saying the data "probably only reflect the inhomogeneous distribution of galaxies in the region of the Rubin–Ford sample," and that "the data . . . are consistent with isotropic expansion, an unperturbed galaxy velocity field and hence a low density Universe."[7] Vera, Judy, and Kent seemed to anticipate that their data and interpretations would possibly spark controversy. "Obviously," they wrote at the end of their paper, "we are not through with this business."[8]

The business was, in fact, far from over; other scientists had become curious about finding evidence for, or against, large-scale structure and large-scale motion in the cosmos. Princeton theoretical astrophysicist Jim Peebles, who had been instrumental in pushing forward the idea that individual galaxies were shrouded in haloes of invisible matter, was one of those scientists and had begun contemplating the structure and movement of groups of galaxies in the late sixties. Peebles was particularly interested in their evolution from the beginning of cosmic time until today, yet by the mid-seventies, astronomers and cosmologists still assumed that the universe appeared the same in all directions, with evenly distributed clusters of galaxies no matter at what angle the cosmos was observed. Such even distribution is what astronomers call an isotropic universe. If there were anomalies in the even distribution, astronomers called those anisotropies. But it was hard to imagine how in an early universe, with such an even distribution of matter, there could be enough gravity and instability in the cosmic media to bring clumps of material together to form stars, galaxies, and ultimately galaxy clusters and superclusters. So Peebles wanted to know: Is the universe truly isotropic after all?

Peebles worked with a student, Daniel Hawley on this question, and the two created an automated, unbiased method to analyze photographic plates to identify isotropy or anisotropy in the universe. A statistical analysis of the orientation of galaxies appeared to show that galaxies do appear to "flop all over the place," Peebles said.

When Hawley delivered his manuscript with this information, part of his PhD thesis, to Peebles in his office one day, John Archibald Wheeler happened to be there. The eminent theoretical physicist had worked with Niels Bohr in explaining the basic principles behind nuclear fission and popularized the term *black hole* for objects that experience gravitational collapse. When Hawley walked in, Wheeler wanted to know all about Hawley's thesis, and when Hawley told Wheeler what he'd found, an isotropic universe with no signs of universal rotation, Wheeler said, "Oh. Gödel will be fascinated by this." Wheeler was referring to mathematician and philosopher Kurt Gödel, who worked at the Institute for Advanced Study in Princeton. Gödel had been working on the rotating universe idea when Vera was doing her master's and doctoral theses in the 1940s and 1950s, and he had suggested, based on mathematics alone, that the universe didn't have a smooth distribution of galaxies; it was anisotropic, and it was rotating. Hawley knew none of this, and when he heard Wheeler mention Gödel, all he said was, "Who is Gödel?"[9]

Wheeler retorted: "To say that Gödel is the greatest logician since Aristotle is to do Gödel a disservice." He then explained Gödel's model of the universe, told Hawley how the mathematician was a towering intellectual figure, and then picked up the phone, dialed Gödel's number at the institute, and handed the phone to Hawley, who melted down, Peebles said. "It was quite a hilarious moment."[10] Despite Hawley's embarrassment, he and Peebles ultimately published a paper in 1975 showing no traces of anisotropy in pairs and clusters of galaxies.[11]

Peebles's simulations, however, indicated that the interactions of galaxy clusters and superclusters with the invisible

matter thought to be surrounding galaxies might also be important for individual galaxy growth. And as data on clusters of
galaxies accumulated, astronomers' calculations began to point
to anisotropies in the cluster galaxies' velocities, which Peebles
wanted to explore. If they were real, they'd offer evidence of
a flat universe, not a saddle-shaped one, shapes that corre-
late to the ultimate fate of the universe. So probing deeper,
Peebles investigated the galaxies' velocities and found that the
anisotropies were not real. They could be chalked up to not
much more than noise in the data. Because of such uncertainties, Peebles's calculations couldn't say much more about the
shape of the universe; either saddle or flat was a "reasonable
possibility."[12]

Vera, however, hadn't given up on her anomalous data.
She, along with Kent, Norbert, radio astronomer Mort Roberts,
and Carnegie astronomer John Graham, collected more galactic velocities and found something very different from what
Peebles had shown. The team was specifically trying to address
the question: Is the expansion of the universe consistent in all
directions (isotropic) or varied (anisotropic)?

If the expansion was isotropic, then galaxies at similar distances from the Milky Way should move away from our galaxy
at similar velocities, no matter their direction. That would mean
the universe is expanding at a steady state in all directions.
When Vera and her collaborators looked at their data, however,
they found the galaxies weren't moving at the same speeds iso-
tropically at all; some even seemed to be speeding toward each
other in a certain region of the sky. There was anisotropy in the
galaxies' velocities, the data suggested.[13] Others have described
the results as an "anisotropy in the expansion of the Universe
on a scale of around 100 million [light-years]." This anomaly
bolstered the claim of the Rubin-Ford effect, offering even

more evidence of some sort of large-scale, anisotropic cosmic expansion. Once again, some astronomers didn't think such anisotropy could exist, and so the team's results, published in the fall of 1976, were met with intense criticism,[14] and even discredited as a statistical artifact.[15] Peebles would later go on to defend Vera's data, writing with Martin Clutton-Brock that they were "reluctant to dismiss a carefully made set of observations even when their implications seem surprising."[16]

A few years after debate about the Rubin-Ford effect emerged, cosmologist George Smoot and colleagues identified a slight anisotropy in the cosmic microwave background, which the scientists initially thought was due to the sun's motion in the galaxy combined with the galaxy's motion through space. This discovery ultimately helped validate some of the large-scale peculiarities Vera and Kent found in galaxies' velocities.[17]

Still, Vera didn't like the negative attention drawn to her work, and she wasn't interested in debating the data, so she turned back to her individual galaxies and their rotation curves. As she'd worked with her galactic velocity data, she'd read the opening line of Peebles's paper with Jeremiah Ostriker and Amos Yahil about each individual galaxy having ten times or more mass than astronomers had originally estimated, an explanation for the galaxies' flat rotation curves. Vera wanted to know if that was in fact right—that is, if many galaxies had flat rotation curves and therefore more mass than we could see. Her curiosity was encouraged by the work of Mort Roberts, Seth Shostak, and other radio astronomers who had collected data on the speeds of hydrogen gas clouds as they spun around their galactic homes, though there were just rotation curves by this point.

There was, however, controversy with these claims too. Responding to the Peebles, Ostriker, and Yahil paper, Vera's former colleague Geoffrey Burbidge argued the exact opposite: that galaxies didn't need invisible halos to stabilize them, so there were not significant amounts of unseen matter lurking in the universe. He cited a 1973 paper by astronomers Darrel Emerson and Jack Baldwin describing Andromeda's rotation curve,[18] which Burbidge alleged slowly declines with distance from the galaxy's center (though that's not evident in the paper itself). The calculations of how much mass there is relative to light, the mass-to-light ratio, is also low, one piece of evidence among many, that gigantic unseen halos around galaxies didn't exist, he contended.[19]

Intrigued, and determined to know whether most galaxies had flat rotation curves, Vera, Kent, Norbert, and a few other astronomers developed a research program in the mid-seventies focused on collecting velocity data on hot, gassy stellar nurseries in lots of galaxies instead of just one or two. The observers started to go galaxy-by-galaxy hunting for these stellar nurseries and using them to draw the galaxies' rotation curves. Vera expanded the types of galaxies she studied too, including prepubescent spirals, with their bright, central bulges but no swirling arms.

Even this kind of stellar city, without a defined swirl structure, had a flat rotation curve. A case in point is NGC 3115, the Spindle galaxy. It sits 32 million light-years from Earth in the direction of the Sextans constellation. As with every other galaxy Vera's team studied, this one also turned out to have stars far from the galaxy's center that orbited the core just as fast as stars closer in. The team did spot an oddity in the galaxy, though: the stars on one side moved around it quickly, at nearly 900 kilometers per second, while the stars on the other side circled much more slowly, at only 400 kilometers per second.[20]

The galaxy clearly had a flat rotation curve, suggesting missing matter. But the differences in velocities on each side of the galaxy were peculiar and hinted that there wasn't as much of whatever was tugging on the stars on one side as on the other side. Now, not only might there be some form of invisible matter in a galaxy, but it might not always be evenly distributed at the galaxy's periphery. It was a prime example of Vera's research testing theorists' ideas. They thought they'd known what galaxies looked like, but this one, while still having a flat rotation curve, threw a wrench in their models. It raised a new question about what was going on in galaxies.

"A truly great scientist will settle questions, to be sure," Peebles said. "But a really, really great scientist will raise more questions than they settle."[21] Vera's look at the Spindle galaxy did just that.

The Spindle galaxy wasn't the only one giving astronomers trouble. Another one, galaxy NGC 1275, was also proving tricky to understand. The galaxy looked like an explosion in the sky, as if a bomb had gone off inside it, leaving fast-moving material at its edges and slower-moving material within it. As far back as the 1950s, astronomers had been trying to figure out why this galaxy was shaped so oddly. Some suggested it was two galaxies that had recently collided. Others said a nuclear explosion in the galaxy explained the fast-moving outer material. Still more said there were actually two galaxies in the image and that they might not be interacting at all.

When Vera and her colleagues took a close look it, they found that the fast-moving material was in fact a galaxy sitting in front of NGC 1275. We see that foreground galaxy edge-on, looking directly into its pancake disk of stars, gas, and dust, while the other galaxy sits behind it. The two-galaxy model that had been suggested before "now rest[ed] on firm observational foundation," Vera's team wrote in early 1977.[22]

It was another galactic mystery solved. This was how Vera worked through the late seventies, going galaxy by galaxy, searching for what made each swirl of stars unique. "Vera got enjoyment out of looking at the individual galaxies. In a way, they take on their own personalities," Norbert said. "Studying them becomes more personal. They each teach you something."[23]

Collectively what they revealed to Vera and her colleagues was that galaxies' rotation curves were flat. By 1978, Vera, Kent, and Norbert published a collection of rotation curves of bright spirals, including data on eight galaxies and two sets of galaxy pairs. All of the stars and gas traveled at similar speeds around the galaxies' cores, even as far out as 160,000 light-years from a galaxy's center. The rotation curves were flat. Mort Roberts and his collaborators had first pointed out a similar phenomenon in the 1960s and 1970s, Vera, Kent, and Norbert had written in the paper, explaining that the radio astronomers "deserve credit for first calling attention to flat rotation curves." By the time the paper came out, Roberts and his collaborators had even more radio data on rotation curves. Coupled with the latest data in visible light, Vera's team wrote, "These results take on added importance" in conjunction with the theorists' suggestion that "galaxies contain massive halos extending to large [radii]." If galaxies did indeed have subluminal halos swaddling them, then Newton's law of gravity worked, and the rotation curves were flat, the results implied.[24]

By the end of the 1970s, the evidence for the existence of dark matter was quickly mounting. Sandra Faber, who hadn't believed there was much to the curves earlier, and James

Figure 12.01
Vera measures spectra of galaxies at the Department of Terrestrial Magnetism.
Credit: Emilio Segre Visual Archives/American Institute of Physics/ Science Photo Library.

Gallagher, an astronomer at University of Illinois, wrote in the *Annual Review of Astronomy and Astrophysics* in 1979 that there were of course uncertainties in the data, possible observational errors, and other issues to be resolved. Yet, they said, "we think it likely that the discovery of invisible matter will endure as one of the major conclusions of modern astronomy."[25]

Even with a consensus growing that supported the idea that invisible matter lurked in galaxies, there were still doubts about galaxies' rotation curves, especially with the radio data, in which noise could make the rotation curves look flat, when they might not be. Vera's work with optical telescopes helped buoy belief that the flat rotation curves were real. A few critics still tried to explain the data away.

Vera pressed on. She described the rotation curve of our Milky Way, suggesting it was flat out to at least 150,000 light-years from the galaxy's center, and she drew out the rotation curve of a galaxy called UGC 2885, one of the largest spirals in the universe. She went on to collect data on nearly two dozen other spirals, specifically focusing in on the Doppler-shifted chemical fingerprints of the stellar nurseries, which revealed the velocities of the gas. She was asking questions and making cosmic observations critical to causing a "paradigm shift" in science, according to physicist and philosopher Thomas Kuhn.[26]

Vera was driving a scientific revolution: her spectral data were key to revealing the ubiquity of galaxies' flat rotation curves and therefore that galaxies had to have dark matter. And it was those chemical fingerprints of a galaxy's young, hot stars that Vera relied on to assuage others' doubts. "You could show someone a couple of spectra," she said, "and they knew the whole story." It was that simple. The spectra showed that the velocities of the stars and gas didn't drop off; the galaxies' rotation curves were flat. Even the behemoth UGC 2885's curve turned over and flattened out, staying constant as far out as the team could detect stars and gas—roughly 200,000 light-years from the spiral's core, Vera reported in the June 1980 *Astrophysical Journal*.[27]

That all of these galaxies had stars and gas that swirled around a galaxy's core at similar speeds was a "genuine discovery," Peebles said. "It was Rubin and Ford who discovered the universality, or near universality of flat rotation curves."[28]

With more and more spectra telling an undeniable story, more and more theorists started to get on board with the idea

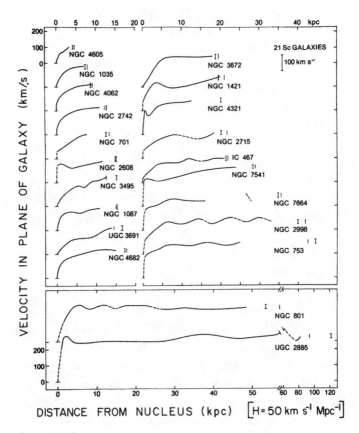

Figure 12.02

Rotation curves for twenty-one spiral galaxies, including the two gargantuan galaxies, UGC 2885 and NGC 801. The curves do not fall down on the right, as expected if only visible matter were there. The curves' flatness—if we understand gravity correctly—implies galaxies have some unseen matter.

Credit: Rubin et al., *Astrophysical Journal* (1980).

that galaxy rotation curves were flat and that dark matter had to exist. This all revealed itself at the December 1980 meeting at the Institute for Astronomy conference in Cambridge. That's why Thonnard hurriedly wrote his postcard from Cambridge exclaiming that finally the astronomy community was starting to believe in the data and its implications.[29] "People understood that if you wanted to save Newton's laws," Vera said, "you had to have lots more matter that wasn't luminous."[30]

13 Diving into Dark Matter

On a temperate November day in New Delhi, Vera was steeling herself to give one of the most momentous talks of her scientific career. She was at the triannual General Assembly of the International Astronomical Union and had been asked to give an "invited discourse," the IAU's version of a top astronomy prize. It was 1985, nearly thirty-five years since she'd given her first professional presentation at the American Astronomical Society meeting in Haverford, Pennsylvania. Back then, Vera hadn't paid much attention to the negative reaction to her work. Now, she was quite comfortable defending her data. Being asked to give this coveted IAU lecture signaled she'd won high status in the astronomy community. As she stood talking to a congregation of astronomers and cosmologists who came to hear her speak, she opened with the idea that nature had been unkind and "played a trick" on all of them. "We thought we were studying the universe," she said. "We now know we were only studying a small fraction of it." That small fraction was the part lit by stars, gas, and dust. The rest of the universe, she said, is made of something else entirely—something "astronomers cannot see in any wavelength," something they call dark matter.[1]

Scientists, she said, were just beginning to probe dark matter, and as a result, they had to ask simple questions, sometimes "deceptively simple questions," about this unseen substance, such as: Where is it? How much of it is there? And what is it? By answering those questions in her talk, Vera said, "I hope the discussion will convince you that dark matter does exist."

By the time she was asked to speak about dark matter at the IAU meeting, many astronomers were on board with the idea that the cosmos contained some unseen substance. Some holdouts still weren't convinced, so as she stood at the podium at the general assembly meeting, Vera set out her intention for her talk: to review the observational data and theoretical framework that made the existence of dark matter all but undeniable.

With little fanfare, she launched into a discussion of the first question she had posed: Where is dark matter? She started by explaining the work she had done with her colleagues studying the velocities of bright emission nebulae, the cosmic nurseries of hot, young stars, as they swung around the centers of bright spiral galaxies. "By analogy with the solar system, in which the orbital velocities are lower for more distant planets, astronomers had long expected that rotational velocities [of stars and gas in galaxies] would first increase with increasing radial distance from the nucleus, reach a maximum, and then fall to low velocities at larger radial distances," she told her audience. In the nearly one hundred bright spirals she had studied, this was not the case. "Falling velocities are not observed. Rotation curves are flat or slightly increasing at large nuclear distances." What's more, she explained, it did not matter what type of spiral she studied; the rotation curves were always flat.

As she spoke, she recognized the contribution of radio astronomers' observations. Mort Roberts and others had clearly shown that gas farther out than any observable stars in a galaxy also whizzed around it way too fast for what Newton's law of gravitation allowed. Dark matter, the data had shown, is hidden in galaxies across the cosmos, specifically at the outer edges of each galaxy. There's more dark matter in a galaxy's outer edge than in any other place in the universe, Vera said.

To further describe where dark matter lurked within a galaxy, she briefly reviewed computer simulations, along with her observations of galaxies that had a ring of matter around their poles. These polar ring galaxies earned only a brief mention in Vera's talk, yet her work on them, as she had told her audience, was essential to understanding how dark matter was distributed in a galaxy.

Though she'd didn't describe her work on polar ring galaxies in detail in her talk, her study of them began in the early eighties when she visited the Cerro Tololo Observatory in Chile. There she'd chatted with Carnegie astronomer François Schweizer, who showed her an image of an odd-looking stellar city that did not have spiral arms but did have an oblong-shaped disk filled with stars, gas, and dust. There was another characteristic feature, too: a ring of stars and gas that didn't orbit in the disk of the galaxy but sweeps around its poles. Probing the movement of these stars around the galaxy, Vera found that the stars probably got stuck at the poles when the galaxy had a close encounter with another of its kind.[2] The data also showed that the stars circled the poles rather fast, meaning dark matter had to be tugging on them too, just as it did on stars in the galaxy's disk. That discovery, along with simulations of galaxy evolution over time, seemed to indicate that dark matter didn't merely circle the disk of a galaxy but instead encapsulated all of it, as if dark matter were like an eggshell around an unbroken egg.

As this evidence about dark matter mounted, astronomers had to concede that a universe without the unseen substance was hard to explain. Vera's work, Princeton theoretical astrophysicist Jeremiah Ostriker said in a phone interview, was crucial to pushing astronomers to accept its existence, and the reason, he said, was that she worked with optical telescopes. Optical telescope data were still more convincing to astronomers than radio data at the time.[3] The radio data did extend the rotation curve of a galaxy much farther than optical observations, and the unfamiliarity of the radio data didn't invalidate all of the radio results; it was just that in the late seventies and early eighties, "to most astronomers, astronomy was optical astronomy," Ostriker explained. "If you found out something using X-rays or radio or something else, that was some other field. But if it was using an optical telescope and using optical light, getting it in the blue or the red," he said, "that was real astronomy." Vera's work, he said, made flat rotation curves and, by extension, the existence of dark matter definitive.

"People then said, 'Wow, this is exciting,'" Ostriker said. "And it was exciting."[4]

Vera's leading role in convincing the astronomy community to believe in dark matter's existence explained why she was asked to give the first IAU invited discourse on the existence of the unseen substance. And so in her talk, after describing where dark matter had been found, she next addressed the question, How much dark matter is there? Roughly estimating the number of stars and the amount of gas and dust in a galaxy allows astronomers to calculate its mass. They can also calculate a galaxy's mass based on the velocities of stars moving around it. Those two methods, however, had resulted in very different answers. The calculation based on the velocity data suggested a galaxy had to have five to ten times more mass than the calculation based on roughly adding up the mass of its stars, gas, and dust.

That much more matter was significant. It meant each galaxy had five to ten times more mass than astronomers had initially anticipated, and taking all the galaxies' masses together and factoring in the additional dark matter that hadn't been accounted for before not only changed the entire composition of the cosmos but also shattered astronomers' confidence in understanding its overall shape and ultimate fate. The addition of dark matter to the composition of cosmos demanded a radical reimagination of the universe, and it drove scientists into wild speculation about what dark matter might be.

Here again in her IAU talk, Vera led her audience through a brief history of what scientists thought dark matter might be, starting with the fond sentiment that it "could be in the form of bricks, or baseballs, or Jupiters, or comets, or mini-black holes," as she'd said in the early eighties.[5] That idea that dark matter was made of stuff we understood was aligned with an earlier guess, batted about as far back as the 1920s by Jan Oort and others, that dark matter was made of very dim stars or gas that telescopes of the time could not yet detect. In the 1970s, radio astronomer Mort Roberts and his collaborator Robert Whitehurst speculated that such invisible matter might be dwarf stars, the coolest and smallest stars in the cosmos. Martin Rees and Simon White of Cambridge University's Institute for Astronomy postulated something similar: that "the most plausible candidates [were] low-mass stars, burnt-out remnants of high-mass stars, or the remnants of supermassive stars."[6] Several years later, some scientists even suggested that dark matter might be made of early universe black holes—regions of space-time so dense, with so much gravity, that nothing can escape them.

The argument that dark matter was some undetectable form of ordinary matter didn't work for everyone, though. Physicists, Vera reminded her audience, had a decent idea of how much normal, or baryonic, matter there was now and how much there had been in the instant after the big bang. According to cosmological theory, the big bang had doled out baryonic matter—specifically, atoms and molecules of hydrogen, helium, and small amounts of lithium—in fairly precise proportions. Adding any ordinary matter to the early universe would make too much helium and not enough deuterium (a hydrogen atom with a proton and neutron in its core, rather than a proton alone). Fiddling with the numbers a bit offered a slim chance that the universe's missing mass could be bound up in failed stars, called brown dwarfs, and dead stars, called white dwarfs. But looking at the data and their cosmological models very closely, astronomers still came up short in their accounting of cosmic matter. Their estimates of the average mass density of the universe, that principal value explaining the shape and fate of the universe, called omega, came out at 0.2, and of that 0.2, less than 0.05 of it was normal, everyday matter. The majority of mass in the universe, at least based on scientists' best estimates, couldn't be baryonic. It had to be something else.

Another set of arguments also bolstered the need for something more exotic than baryonic matter. Those arguments stemmed from unanswered questions about why the relic radiation of the big bang appeared smooth, especially when the universe around us seemed so clumpy. One response to this conundrum came in the late seventies and early eighties when Alan Guth, Alexei Starobinsky, and Andrei Linde postulated a theory called inflation. Inflation is a period at the start of the universe that lasted literally an infinitesimal fraction of a second after the big bang. At that time, the universe was riddled with quantum fluctuations, tiny gravity wells that contained the seeds of galaxies. When the universe expanded exponentially fast during inflation, those quantum fluctuations were magnified, allowing for mass to accumulate into stars, galaxies to grow, and long, wispy tendrils of galaxies and galaxy clusters to take shape.

The theory of inflation therefore explained the origin of large-scale structure in the cosmos and the inconsistencies in the smoothness of the big bang's relic radiation. The theory also required that dark matter needed to be something much more bizarre than baryonic matter because the theory predicts that the cosmos is flat, so that the critical density of omega equals 1. The only way to get there is to have dark matter in a form far different from any type of matter we have detected thus far.

An argument for dark matter's exotic nature had been circulating in the scientific literature as early as 1974, and by 1977, there was some speculation on what this nonbaryonic dark matter might be. But it took another five years, and a definitive acceptance of the existence of dark matter, for these guesses to begin to captivate the attention of the scientific community.[7]

Once the astronomy community accepted the existence of dark matter, physicists went wild, dreaming up all sorts of dark matter candidates. In return, cosmologists used their calculations about the birth and evolution of the universe to put limits on those hypothetical particles. Up first was the neutrino. It took on two forms, one svelte and another heftier. Serious flaws arose fairly quickly with svelte neutrinos, and it was Sandra Faber, who'd worked with Vera in the 1970s and wrote the paper proclaiming scientists take a hard look at the evidence for the existence of dark matter, who knocked them out of the running as *the* dark matter particle. Her observations of small, spheroidal galaxies showed dark matter particles had to have more mass than what was calculated for the light neutrino, and after her results were reported, the poor little particle was out of the dark matter pool.[8]

Particle physicist weren't deterred. They next turned their attention to the heftier neutrino. Its hypothetical existence was invoked in the late seventies to make omega more than 1 and "close" the universe, giving the cosmos enough mass to overcome expansion driven by the big bang. In this scenario, the universe would recollapse and ultimately implode.[9] Reading about this proposed particle in 1978, astronomer James Gunn, then at Caltech, and Gary Steigman, at Yale, saw it as a perfect fit for the invisible mass Vera and others had found in galaxies. There could not be better stuff to constitute dark matter than these neutrinos, they wrote. In the early universe, these particles would bind together, creating gravity wells that would attract regular matter, allowing stars and galaxies, with unseen halos, to grow, according to the theory.[10] But Faber's analysis put the general idea of dark matter being made of neutrinos on shaky ground, so physicists came up with other potential dark matter particles.

Steigman and Michael Turner were advocates of a creative culprit called "a weakly interacting massive particle" (WIMP). WIMPs held so much promise as a dark matter candidate that particle physicists launched efforts to build detectors to snare one. As they did, other physicists suggested the axion as a dark matter particle. The axion was first described in the late 1970s and named in honor of a laundry whitener, as it was supposed to "clean up" some messy details in particle physics.[11] By the 1980s, axions were regarded as a particle that "fits the bill remarkably well" for dark matter, James Ipser and Pierre Sikivie wrote.[12] And so again, as they did with WIMPs, physicists thought up elaborate experiments to see if axions did indeed exist.

With neutrinos, WIMPs, axions, and other alternatives, the pool of dark matter candidates was turning into quite a particle zoo. In her IAU talk, Vera briefly reviewed all of the candidates, then unflinchingly addressed what might have been an elephant in the room at the time: the idea that dark matter didn't exist at all. Instead, some scientists suggested, a galaxy's flat rotation curve could be explained with a modified theory of gravity. Mordehai Milgrom, an Israeli physicist, suggested it was in fact a bit shortsighted to use Newton's laws, strictly as they were written in the seventeenth century, to work out the dynamics of galaxies. While these laws would be reliable when dealing with, say, the solar system or an apple falling to the Earth, they may actually break down at the scale of galaxies and the vastness of space, Milgrom argued. He decided to modify both Newton's laws and how they worked at small accelerations typical for galaxies. Milgrom's solution, with Jacob Bekenstein, to explain a galaxy's flat rotation curve was called MOND (modified Newtonian dynamics). And, surprisingly, the adjusted arithmetic was able to reconstruct the curves' signature flatness.[13] Not many scientists took stock in such a radical reassessment of Newton's laws on the galactic scale or of dark matter's existence.

Vera, however, kept an open mind. Faithful to her motto of never assuming something is true just because lots of people say it is so, she said she wished the MOND proponents were at the IAU. "I am sorry they are not here to discuss their work," she told her audience. Hearing from them would only have been fair, she seemed to imply. She let that thought sink in, then moved on to remind listeners that the existence of dark matter "was a story," a story whose details we are just beginning to comprehend. Her talk—a summary of the state of astronomy, cosmology, and particle physics as it related to dark matter in the mid-eighties—made clear that some questions about universe had been answered by her work and the research of others. But "many of the answers still elude us," she said. "We have a lot to learn."[14]

14 Gender Equality in Astronomy:
A Darker Universe?

Just as her observing runs and speeches at meetings far and wide kept Vera away from her family, sometimes it was her family who reminded her of her duty to astronomy and to the other women who wanted to become astronomers. On an evening in 1975, Vera, Bob, and their youngest son, Allan, were sitting at the family's long, wooden dining room table—the one Vera often worked at when she was wasn't in the office. Together, the family was discussing an invitation Vera had received to give a lecture at the prestigious Cosmos Club in Washington, DC. She'd be speaking to the Washington Philosophical Society, a science education organization that regularly invited researchers from around the world to share their latest work. Vera was honored by the invitation. But there was one condition for giving the talk that made her uneasy: women weren't allowed to enter the club through the front door, so even though she was the guest lecturer, she'd have to enter the club through the side door. She now faced a difficult decision: go in through the side, unseen, and present her work on galaxies, or do not go at all.

Choosing what to do wasn't easy. The Cosmos Club had been one of Vera's first connections to astronomy way back when she was in high school and the Goldbergs took her there to see Donald Menzel speak. She wanted to present her work there. But she didn't want to enter through the side door, a signal that she was somehow second rate because she was a woman. Allan, a teenager at the time, offered a solution: "Walk through the front door and say [you are] a man," he told her. His mother didn't take her son's advice; she walked in through the side door as she was asked to do. "I was surprised," Allan recalled later. "That was uncharacteristic of her to back down like that."[1]

It was one of the rare times Vera would ever back down on account of her gender. She'd been increasingly more vocal about disparities for women in science, especially as she heard anecdotally from her daughter that encouragement of women in the field was no better than when Vera herself had been a student. Vera recalled that when Judith was at Radcliffe, "her very young, very bright advisor suggested she go off and get married the first time she went to him with a problem."[2] Not long after her daughter told her the horrid story, Vera worked with fellow astronomers Anne Cowley, Roberta Humphreys, and Beverly Lynds on a report for the American Astronomical Society (AAS) that summarized the status of women in the field. In the early 1970s, the AAS had formed a group to gather data and anecdotal stories of women in astronomy and document the challenges they faced searching for jobs, conducting research, and earning astronomy degrees. When the group collated the data, they found a startling trend: in the 1970s, the society had the smallest percentage of women it had ever had. In addition, it had never had a woman president. It had had only two female vice presidents compared with fifty-eight men in the role, and women had never solely won an AAS award that accepted applications from both men and women. (Margaret Burbidge had shared a prize with her husband in

1959, but there were no other female winners.) The only prize women won was the Annie Jump Cannon award, which was specifically given to women. Margaret Burbidge had been named as the recipient, but in 1972, she refused it on the basis that it was solely for women. In 1974, the organization that awarded the prize shifted; instead of coming from the AAS, it came from the American Association of University Women with advice from the AAS. Elimination of the sex-restricted prizes led the AAS report authors to wonder: Would that increase the probability of women being considered for other awards—awards men typically won?

Aside from awards, women's faculty appointments were also drastically lower than men's, and most women with astronomy PhDs were in less prestigious, untenured positions compared with the male AAS members, probably because of sexual discrimination in hiring, whether blatant or not, Vera, Cowley, Humphreys, and Lynds reported. The situation was slowly changing, with the largest telescopes now open to women and graduate schools that had previously excluded women—Princeton, for example—now accepting them. Still, other issues continued to stymy women's progress in the field. A pay gap existed between men and women with the same qualifications, and lower pay for married women was even considered justified "by the reasoning that since her husband works she doesn't really need the money." Added to that were differences in access to publishing papers in high-quality journals, though women had the same productivity (about four papers a year) as men. And the percentage of women receiving astronomy PhDs fell; from 1947 to 1952, it was 12 percent, but from 1962 to 1972, it was only 8 percent, the group found.[3]

Based on the data, the team wrote, "We believe that women face far greater obstacles in almost all aspects of their professional careers than do their male counterparts." And, they noted, the astronomy community would be greatly enriched if women were accepted and employed as equal colleagues. They pushed for a list of female AAS members to be circulated widely within the society so that women could be considered for society officer positions, appointment to AAS committees, paper presentations, nomination for prizes, and many other facets of society inclusion. They also urged universities to adhere to affirmative action in hiring, rewrite nepotism laws that might prevent a woman from getting a faculty appointment in the same department as her husband, and watch for conflict of interest instead. Finally, they requested that the president of the AAS send their report, with its recommendations, and a letter of support to department chairs, observatories, and other places that employ astronomers to encourage them to follow the recommendations.[4] There's no clear record of whether the AAS president at the time, Robert Kraft, heeded the recommendations of Vera and her colleagues. But Margaret Burbidge did become AAS president in 1976, after Kraft's tenure was up, and during her presidency, she convinced AAS members to ban meetings in states that had not ratified the equal rights amendment to the US Constitution.[5] So it seemed that Vera's first formal foray into advocating for equity among men and women in astronomy had some effect. It wouldn't be her last time speaking out.

It is curious, then, that a year after finishing the AAS report, when given the opportunity to defy convention at the Cosmos Club, she yielded. Perhaps it was a battle not worth fighting. She'd also been warned that advocating strongly for women could affect her career. It was these events that set the stage for Vera to use her voice repeatedly to speak out about sexism, especially in science.

The same year Vera yielded at the Cosmos Club, Judy, her daughter, defied the recommendation of her advisers. She was then in her early twenties, enrolled as a PhD student in astronomy at the University of Minnesota, and was about to be married, with the conviction that marriage would not end her astronomy career. In response to the news, the faculty suggested she terminate her degree after earning her master's because, as Vera recalled, Judy "could not be very serious about being an astronomer if she were getting married." Judy was serious about both, so she got married and continued her astronomy career. Whether a man would have faced similar criticism because he chose to get married was hard to say, Vera later reflected, but women did seem to face many more obstacles.[6]

One of those obstacles, Vera said, was the language used in science itself. In 1978, she penned a letter to the editor of *Physics Today* noting that while efforts to include capable women in research were improving, the language used reflected that science was a profession dominated by men. She wanted the language to be more inclusive, to demonstrate that anyone, regardless of gender, could do science. She was taking particular aim at forms, proposal review requests, and NASA applications that consisted of questions about the applicant's qualifications, specifically what *he* accomplished, or what *his* data showed. Written that way, the documents implicitly suggested science was done only by men, Vera argued. If the appearance of science is to be inclusionary, not exclusionary, of women, she said, then these kinds of "changes in language may have to lead the way."[7]

She'd again call out sexism in science and science writing in *Physics Today* a few years later, this time pointing out that an article instructing researchers on how to give lectures on their work was particularly tone deaf to the fact that women were physicists too. All of the instructions, Vera noted, were directed toward men. Most grievous was the case that when a female pronoun was included, it referred to a dancer falling on her face. "In the male world of physics . . . presumably only females fall on their faces," she wrote, criticizing the piece and the editorial comment that accompanied it, especially for claiming the advice was as relevant today as it was when it was first published. (The piece ran in the magazine in 1951, 1961, and again in 1981.) "At least the falling dancer could have been male," Vera quipped in her 1982 response.[8] It was a year after Vera had been inducted into the National Academy of Sciences, another place where she'd tried to push for the advancement for women in science. When she stood to talk at academy meetings, her sister recalled, the leaders groaned because they knew she was going to ask: What are you doing about getting women into the academy?[9]

Attitudes were continuing to shift, though at a glacial pace. The same year that Vera wrote that male physicists never seemed to fall on their faces, Margaret Burbidge was elected president of the American Association for the Advancement of Science, the world's largest general scientific society. In 1984, Burbidge earned the highest honor of the AAS, the Henry Norris Russell Lectureship, and a year later, the same year Vera gave her invited discourse at the International Astronomical Union meeting, Sandra Faber won the AAS and American Institute of Physics' Dannie Heineman Prize for Astrophysics, which had previously been awarded only to men. Perhaps most encouraging was that women were observing at telescopes around the world much more frequently; a group even sent a cable to Vera while they were working, telling her excitedly that there were only women observing at all the large Cerro Tololo telescopes in Chile one evening.[10] It was a small step toward equity.

While Vera advocated for better representation and recognition for women in science, she continued to go to the telescope to probe the behavior of galaxies. Because of her fascination with galaxies of all types, the standard and the bizarre, she and her colleagues expanded their sample, adding objects like the behemoth UGC 12591, a massive swirl of stars that sits about 400 million light-years away from the Earth in the direction of the constellation Pegasus. When Vera and her colleagues tracked the velocities of stars as they shot around this galaxy, they found something surprising: not only were the outer stars moving at the same velocities as the stars closer in, but all the stars seemed super-speedy. They whizzed around the galaxy at 500 kilometers per second—roughly a million miles per hour, two times faster than the orbital speeds of stars in the Milky Way, meaning this galaxy had to have massive amounts of dark matter. Vera was left to wonder: Was this galactic beast on its own out there in the universe, or were there other supermassive spirals like it elsewhere?[11]

As she searched for exotic spirals to study, Vera also started working with astronomer Deidre Hunter, who had joined the Department of Terrestrial Magnetism as a postdoctoral fellow. Hunter had earned her PhD in astronomy in 1982 and worked as a postdoc at Kitt Peak National Observatory for a few years before coming to Carnegie. Together, Vera and Deidre, with help from John Gallagher III, who was on staff at Kitt Peak then, began to hunt for stars and hot gas farther and farther from a galaxy's center. One of their first galaxy targets was a spiral with a bar at its core. Called NGC 3198, this galaxy sits 47 million light-years from Earth in the constellation Ursa

Major. When Vera, Deidre, and Gallagher examined it with the 4-meter telescope at Kitt Peak, they found evidence of star formation near the very fringe of the stellar swirl's pancake-shaped disk. It was a rare find, and they needed more data to determine exactly was happening out there, but, they concluded, it was a first peek in optical wavelengths at what goes on with gas and stars at the very periphery of galaxies.[12] The discovery would drive them to hunt for stars farther and farther out at the edges of other galaxies, including supermassive ones such as UGC 2885.

Though focusing mainly on individual galaxies, Vera did return to the question she'd asked in her master's thesis and again with Kent Ford and her daughter, Judith, in the 1970s. The question now was: Is there large-scale, peculiar motion of galaxies outside of what was predicted by the Hubble constant—the velocity at which galaxies were moving away from one another because of cosmic expansion? On their way to investigating this question, Vera and fellow Carnegie scientist Norbert Thonnard described a new way to calculate the Hubble constant using galaxy rotation curves. In the eighties, one group estimated that the constant was roughly 50 kilometers per second per megaparsec. (A megaparsec is roughly 3.26 million light-years.) Not everyone agreed, however. Another group estimated the constant to be as high as 90 kilometers per second per megaparsec. From their data, Vera and Kent Ford could also deduce an estimate for the Hubble constant.

"The results are surprising," they wrote. Their data crunching suggested the Hubble constant was roughly 80 kilometers per second per megaparsec (Mpc). Their work showed that the lower estimate held by some astronomers at the time was possibly wrong. (The current estimate is still up for debate, with some data showing a Hubble constant is 67.4 km/s/Mpc, while other estimates have it at 73 km/s/Mpc.) In addition, the method the pair devised offered another way to calculate the actual brightness of galaxies, and, in turn, estimate more precisely how far each one was from Earth.[13]

Drawing on that work, Vera looked for random motions beyond the expected spread of galaxies in her sample of spirals and the galaxy clusters in which they resided. Again, she found unexpected movements.[14] Sandra Faber and her colleagues had also shown unexpected velocities for elliptical galaxies, and together, the results raised "severe cosmological questions" about gravity: How could it allow for these peculiar, large-scale motions across such vast regions of space, especially if the universe was homogeneous throughout?

An extensive survey of galaxies, done at the time by a team at the Harvard/Smithsonian Center for Astrophysics that was headed by Margaret Geller and John Huchra, revealed that stars and galaxies are distributed on the surfaces of vast intersecting bubbles, while the voids filling the space within them were quite empty. It was a revolutionary discovery that, if correct, would mean astronomers would have to revise their models of cosmic evolution all over again.[15] Collectively, the work was a clue that perhaps scientists again didn't understand large-scale structure and motion of galaxies in the universe as we see it now or even at the beginning of cosmic time, at least not as well as they had thought. What the data did reveal was that the universe, Vera wrote in 1988, is distributed in a "non-random, clumpy manner," and that clumpiness "gives rise to large-scale bulk motions," which "lacking evidence to the contrary . . . can be described by Newtonian gravitational theory."[16] In other words, her graduate work done some three decades before had been partially right: there was large-scale structure and motion in the universe, though maybe not a universal rotation.

Not long after these discussions started circulating in the literature in the late eighties, more data would buoy the idea of clumpiness and empty space, cosmic bubbles and voids. This time, however, the data didn't come from astronomers studying galaxies with telescopes on the ground. It came from a precise look at the cosmic microwave background, the relic radiation, of the big bang. The data were from COBE (the cosmic background explorer), a satellite launched into space in 1989 to take precise measurements of the universe's relic radiation. When the satellite beamed its data back to Earth, the precision maps of this relic radiation stunned astronomers. They could clearly see that there was a background signal at 2.725 Kelvins, confirming that the early universe was hot and dense and then quickly expanded and cooled. The signal was largely homogeneous, though on careful inspection, there was evidence of small anisotropies, tiny temperature fluctuations caused by small variations in gravity in the early universes. Those tiny fluctuations provided the evidence that matter would eventually clump into clouds of gas and then into stars. And it was just those variations in the cosmic background that could give rise to the wispy web of galaxies spooling throughout the universe.

Even with that exciting discovery, though, the COBE results presented a small problem: the calculations COBE gave for the amounts of dark matter and normal matter still didn't get cosmologists to their desired cosmic matter density of omega equals 1. Cosmological theory suggested that omega should be 1 and that the universe should be flat, but COBE seemed to say this wasn't the case.

Once again, not all cosmic questions were answered, and many more were raised.

Vera remained focused on her galaxies, setting off to observe at the illustrious Palomar Observatory to study them, and while she was there, she couldn't refuse the opportunity to remind the astronomers of gender discrimination in science. As she prepared for an evening of observing with the famous 5-meter telescope, she was indoctrinated into the new ritual of observing from the warm, computer-operated observing room rather than the cold dome and cage where she used to take observations on photographic plates. She did walk into the telescope dome, though, and instantly saw the historic bathroom only for men. That's when she got a little cheeky, and drew a skirted woman, then pasted it on that bathroom's door. The sign stayed up for the four days she was there, but it was gone again when she returned a year later.

Vera was not offended by the sign being taken down. Her point had been made, and truly, for her, by this time, she said, her focus had shifted from breaking new ground in astronomy to focusing more on her family. Her children had started having children, and Vera wanted to "recreate her childhood memories with sweet grandmotherly aromas" for her own grandchildren. Working constantly, she worried, would make her too busy to be the grandmother she'd remembered having as a child, so she found herself often choosing to go see her grandchildren rather than traveling to yet another scientific meeting.[17]

To foster family time, Bob and Vera bought a condo in Jackson Hole in the late eighties, "on the philosophy that if they wanted to entice children and grandchildren from all over the country then they had to pick a nice spot," Allan, their son, said. The couple's children and their families were spread across the country, with Judy working as an astronomer at the University of Massachusetts Amherst, Allan as a geologist at Stanford, David a geologist with the US Geological Survey, and Karl as a mathematician at Ohio State University. Yet the planned worked.[18] Every year, some mix of children and grandchildren visited Vera and Bob in Wyoming.

With the family gatherings becoming a priority, Vera again found herself learning to balance work, family, and her social life as she'd done decades ago as a wife and mother. In one particular week, she recalled, she visited Vassar in New York to honor the school's president; on a Thursday, she spent some time with her children and grandchildren who were close by; and then by Sunday, she was back in DC to attend a National Academy of Sciences dinner for a Russian dissident, Andrei Sakharov. Astronomy, she admitted, had become an extra family member, one that could be fun but sometimes exasperating.

That shift in priorities, however, didn't stop her from going to Palomar when she was given observing time, and it didn't stop her from stunts, such as the bathroom cutout, that served as a reminder of the inequities that women in astronomy suffered. That deed represented the tip of the iceberg of her activism, which became more emboldened as she advanced in her career. She again started speaking out more fervently about equity for women in astronomy, giving talks, sometimes with her daughter, at different venues explaining the state of women in the profession and how things had and hadn't changed over the course of her career.

In one of these instances, she described how her life and work rested on three basic assumptions:

1. There is no problem in science that can be solved by a man that cannot be solved by a woman.

2. Worldwide, half the brains are in women.

3. We all need permission to do science, but for some reasons, that are deeply ingrained in history, this permission is more often given to men than women.[19]

These assumptions drove Vera to advocate for herself and other scientists who faced discrimination. In the course of her career, she had collected harrowing stories of how women were treated by their male colleagues, who were chauvinist, dismissive, even elitist in their sexism. They would say, "You do science because we let you," a direct quote from a man who later became a leader of an American scientific society.

"It is hard to know whether to laugh or cry at these tales," Vera said.

She told and retold these tales because little girls needed role models to show them they could be astronomers and other types of scientists, she said. Older girls and college women needed those role models too. And women in science, whether four, forty-four, or older also needed to be themselves, Vera wrote in one of her many essays. "There are enormously diverse styles of doing science," she said, "and the variety will increase as science becomes more egalitarian."[20]

She was fierce with her criticism, when it was warranted, and she protected her family time as her career wore on; and though she had become more family focused and driven in her advocacy, she hadn't completely forgotten her galaxies. With Hunter, she studied how swirls of stars evolved in clusters and how that was different from the way they evolved if they didn't have neighbors nearby. With others, she identified a galaxy, NGC 4550, that had stars circling in both directions, clockwise and counterclockwise, around the core of the stellar city. It was a rare sight, one that probably evolved with time, Vera suggested. Here's what could have happened: As the galaxy took shape, an initial set of stars formed and started circling it, going one way. Then, about a billion years ago, the galaxy drew in a substantial amount of gas, and that gas was the perfect place for stars to form, even though they moved the opposite way around the galaxy's core. Theorists had actually modeled these kinds of galaxies decades before, and so, Vera said, her team's observation of a two-way galaxy made those predictions much more than "elegant curiosities."[21]

Her work eventually earned her major recognition. In 1993, she was awarded the National Medal of Science for "her pioneering research programs in observational cosmology, which demonstrated that much of the matter in the universe is dark, and for significant contributions to the realization that the universe is more complex and more mysterious than had been imagined."[22] That same year she won the Dickson Prize in Science, and in 1996 she received the Gold Medal of the Royal Astronomical Society. A year later, she published a collection of essays she'd written, along with talks and interviews she'd given, in what became an autobiography of sorts, titled *Bright Galaxies, Dark Matters*. Capturing Vera's fascination with galaxies' personalities and idiosyncrasies, the book reveals how she helped to gather irrefutable evidence for the idea that "most of the matter in the universe is dark."[23]

What Vera didn't expound on in the book was how her research served as one of many sparks to ignite the flames that have fueled the hunt for the dark matter. After astronomers, cosmologists, and theorists largely agreed that dark matter had to exist, teams of researchers started to plan searches to indirectly detect it with space telescopes or directly capture it, if it were, in fact, a particle. They dug deep into the Earth, designing elaborate detectors to snare a dark matter particle.

Of course, what no one anticipated, not even Vera, was that the universe was far darker than we could have ever imagined.

This realization came not long after Vera had earned many of the awards for her work related to dark matter.

The discovery that the universe was even darker was made in the late 1990s. And it was made by accident, as two groups of astronomers started to intently study a particular class of supernova. In this class of supernovas, two stars circle each other. Typically, one is a white dwarf, and the other is a star of any kind, as long as it has an extended atmosphere that can be tidally ripped off by the white dwarf's gravitational pull. When the white dwarf gathers in enough gas from the other star, it explodes. The explosion is what scientists call a Type 1a supernova.

Astronomers like to study these kinds of explosions because they all have the same intrinsic brightness. It's possible then to use the observed brightness differences to determine the supernova's distance from Earth. This gave astronomers a powerful new tool to measure cosmic distances. The two teams decided to hunt for extremely distant supernovas, and the galaxies they were part of, to refine the Hubble constant and check whether the rate of expansion of the universe was slowing down, speeding up, or constant? If the cosmos was decelerating, which astronomers assumed, that meant that the combined masses of all the stars, galaxies, dark matter, and the other stuff in the universe would overcome expansion in the wake of the big bang and pull the cosmos back. But that's not what the two teams found. The supernova measurements told them the universe's expansion was doing the exact opposite: galaxies were flying farther and farther apart at faster and faster speeds so that the rate of expansion was accelerating.[24] Something was pushing the universe apart. Riffing off of astronomers' use of the term *dark matter*, Michael Turner decided to call this mysterious force *dark energy*.[25]

This new and unexpected finding reawakened a long-standing cosmological puzzle. When Albert Einstein created his field equations for general relativity, well before Hubble's observations in 1929 but at a time when other theorists were speculating about the fate of the universe, he added a constant to make sure his model described a static universe, not one expanding or contracting. His "cosmological constant" used the mass of empty space, or "vacuum energy," to keep the cosmos stable. After Hubble's observations and confirmation in the early 1930s that the universe was indeed expanding, it negated the need for the cosmological constant. Einstein was embarrassed by this, calling his failure to predict a dynamic universe his "greatest blunder."[26] The supernova work in the late 1990s, however, suggested Einstein maybe had not made a blunder at all, but had in fact described dark energy.

Some mysterious force came up again when astronomers introduced the idea of inflation—the rapid expansion of the universe after the big bang. That theory predicted that something like "negative pressure," similar to the idea of dark energy, would create a repulsive gravitational field and push the cosmos apart quickly in the instant after it was born. That dark energy, however, would have disappeared by the time the universe was just a fraction of a second old. So scientists weren't sure that the negative pressure of inflation was related to what now seemed to be accelerating the expansion of the universe. Like dark matter, there had been clues that dark energy also existed. Now the evidence, similar to what was seen with dark matter, seemed all but irrefutable.

And there was something else that was intriguing about it: dark energy could bump up that value of omega. Based on the new data, a reevaluation of the composition of the universe revealed that matter and dark matter would make up roughly 30 percent of its mass-energy content; dark energy would make up the other 70 percent. Astronomers had finally gotten their critical density of omega equals 1, complete with a flat universe. With dark energy, however, the fate of the universe would change. It would no longer expand forever, at an incessantly decelerating rate; rather, as a result of constant acceleration, it would flicker out and possibly even tear apart in a Big Rip.

This work was enticing, and it touched on supernova explosions and cosmology, all of which Vera had dabbled with before. Still, Vera kept her focus on dark matter, specifically where it situated itself within galaxies and what it did to the stars and gas at galaxies' outer edges. Knowing more about that was what drove her to telescopes time and again. It is why she braved the chill of a November night, at the age of nearly eighty, to return to Kitt Peak to again search for stars and hot gas at the very fringes of gargantuan galaxies. She wanted to know: Does dark matter taper off at the edge of galaxies' spherical halos, or does it pervade intergalactic space?

15 The Final Nights

Ominous gray clouds covered the evening's sky at Kitt Peak. It was a cool November night in 2007, and it was hard to see the stars. Vera tugged at her coat zipper. "Maybe it will get better as the temperature drops," she said to Deidre Hunter.[1] Vera seemed fatigued, and she was distracted.

A few hours earlier, she had called Bob. For the past eight years, he'd been battling a debilitating form of bone cancer. During much of that time, Vera had stayed by his side. Choosing to forgo observing runs, she would instead analyze her data and do her research only at the Department of Terrestrial Magnetism, which was just a few blocks from their home in Northwest Washington. That way, Vera said, she could get to Bob quickly if he needed her.

During the previous few months, from about September on, Bob had been doing quite well, she said, and his health seemed to be improving. By November, she felt confident she could leave him for two weeks to come out to Kitt Peak and study galaxies UGC 2885 and NGC 801. Now, Bob had told her he was suffering severe pain. Vera, thousands of miles from home, could do very little to care for him. She was disheartened.

To divert her attention, she checked the weather forecast. The next two nights were her last two in the fourteen-night observing run with Hunter. The forecast: cloudy tonight; partly cloudy tomorrow. Perhaps she should leave in the morning, she muttered aloud.

Silence.

Vera's eyes closed. After a few moments, she began to reminisce about the days when she felt like a pioneer, when she literally looked up into the sky, fought the cold, and guided the telescope with her bare hands. She vividly remembered that first time at Kitt Peak, nearly half a century ago. Peering through the eyepiece of the 36-inch telescope, she was tracking stars as they traversed the sky and gathering data on the Milky Way. Former observatory director Nicholas Mayall came to check on her, she said. She told him just how wonderful it was to be out there. Without hesitation, he replied, "That is what happens when you pour a million dollars onto a mountain top."[2]

Another beep.

Vera's eyes shifted to Hunter's computer screen where the swirling giant UGC 2885 flickered onto the screen. This galaxy, one of the largest spirals known, is ten times bigger than the Milky Way. At a breadth of 800,000 light-years, the stars in the outermost regions have undergone a mere seven revolutions since the beginning of cosmic time. In that time, our own galaxy has rotated some fifty times. Despite so few complete turns, the spiral arms of UGC 2885 are startlingly smooth and well developed, Vera said, her voice carrying a hint of excitement.[3]

The last time she studied this galaxy, twenty-some years ago, she was startled to find such perfection in the arms. She questioned how such symmetry could happen. It could obviously not be a result of multiple rotations. That she knew. Perhaps it has something to do with the stars of the spiral.

The cradles of gas holding the galaxy's nascent stars circle the swirl's outer edges at similar or somewhat quicker speeds than the stars circulating closer to the center, just as in many of the other galaxies Vera studied.

These curious characteristics are what kept Vera coming back to Kitt Peak. But there was something else too: she returned because she said she was never sure how many more times she would get to study the heavens through the eye of a telescope.

She said she'd hoped to better explain her beloved galaxies' shape, formation, and evolution. To her, it seemed so terribly obvious to keep pushing the limits. New detectors and cameras made her think it was possible to find a galaxy's most remote stellar nurseries, her beloved H II regions, the nurseries of hot, young stars. Finding them, she said, would let her draw out the galaxy's rotation curve to a distance up to three times greater than what was known then, telling astronomers more about the invisible mass that tugged the stellar cocoons along.

Even then, in 2007, scientists still weren't sure what dark matter was. The first guess, as Vera had explained, had been neutrinos. But it turned out that neutrinos were too light to create the dark weight of the cosmos. They were not heavy enough to stir up Vera's stars; plus, the neutrinos flying around as relics of the big bang could not have brought about the bubbly, large-scale structures of galaxies that saturate the universe today. Neutrinos were out.

Next up: axions and weakly interactive massive particles—WIMPs. Both of these subatomic creatures should interact very rarely with normal matter, and the big bang might have produced enough of them to serve as dark matter. But by the time Vera was at the telescope in 2007, dark matter detectors hadn't snagged a positive signal of an axion. And no one had convincingly caught a WIMP, although Italian physicists claimed they

Figure 15.01
Vera sits in her office at the Department of Terrestrial Magnetism during an oral history interview with the author (left).
Credit: Smithsonian Institution.

had detected them. (Other researchers did not widely accept that detection.) Nevertheless, physicists carried on designing and building new experiments every year to look for dark matter particles, and they still are hunting for them now.

With so many instruments and so many professional careers invested in this new particle physics venture, Vera said, "That's a faucet that will be hard to shut off."[4] Her seemingly "dull" investigation of individual of galaxies was among the first to turn on the nozzle, proving her project started in the sixties not so boring after all. Her observations ultimately helped to convince scientists that dark matter is what holds galaxies together and allows them to form in the first place. The galaxies are what let astronomers assume dark matter is there—if it is, in fact, there.

Data released in 2006 of the Bullet Cluster—the remnants of a violent collision of two large clusters of galaxies—offered fairly strong evidence that dark matter does in fact exist. When the two galaxy clusters collided, it was an energetic event, second in strength only to the big bang itself. The normal matter from each cluster slammed into each other, slowing its movement as a drag force—a force similar to air resistance—tugged on it. Dark matter didn't slow down at all. Because it does not interact with normal matter or itself, it cruised on through the collision, creating a distinct separation of the two types of matter in a famous composite image with data from the Chandra X-ray observatory, the Magellan telescope in Chile, and the Hubble Space Telescope. In the image are two pink clumps, one from each galaxy cluster lying side-by-side; that's the normal matter. Each pink clump is flanked by a splotch of blue, the dark matter, which moved through the cluster much faster than the normal matter. Such separation provides the best evidence to date for the existence of dark matter.[5] Yet even at the telescope in 2007, Vera did not insist that there had to be dark matter in galaxies. She thought it made sense to consider other possibilities, such as modified Newtonian dynamics (MOND).

MOND believers of course continued their work, just as physicists pursued their dark matter particles and astronomers kept pushing new technology to its max to determine how dark matter distributes itself in the cosmos. Vera was among those astronomers.

After this long observing run with Deidre, Vera returned home to find her husband in terrible pain, his bones deteriorating after years battling multiple myeloma. As the weeks passed, his health continued to decline.

Bob died in January 2008.

Vera was heartbroken. She'd lost the love of her life, her greatest champion, her best friend, a man who'd always thought about his career alongside her own, often second to her own, Vera's sister, Ruth Burg, recalled. He was soft-spoken, humble, and calm, someone who'd talk about rational and irrational numbers while walking the couple's daughter to middle school, someone who'd clandestinely rendezvoused during the Cold War with refuseniks (Soviet Union Jews barred from emigrating to Israel) to deliver birth control pills.[6] He'd instantly fallen in love with Vera the moment he met her and loved her his entire life, and she loved him just the same, from the start of that family dinner in the 1940s. And now he was gone.

Vera retreated inward; she kept her feelings inside, rarely discussing her husband's death, even with her sister. It was not because she didn't feel the loss, Burg said, but because she felt it so unbearably. "Her own deterioration began when Bob died," Burg said. "She could never really get over it."[7]

Vera did, however, keep living. That's all she could do, and that's all she knew to do; it was the example her father had set when her mother died. Key to keeping her feelings locked away was to keep busy, so Vera agreed to give a talk at the Cosmos Club (where she could now enter through the front door), participate in the World Science Festival, and return once more to Lowell Observatory with Hunter to study the stars. There, she didn't just watch the heavens but shared her love of the stars, helping Hunter show students from the local Hopi Mesa School how to use star charts to find the celestial objects in the night sky. "After dark, the class met again to view the sky, first with their eyes, next with their charts, then with binoculars, and finally with the telescope. There was little silence, some confusion, lots of questions, and some shouts of excitement," Vera recalled. "Before the night was over, one of the very bright students announced his interest in becoming an astronomer."

Vera, Deidre, and other astronomers from Lowell also worked closely with some twenty Native American teachers, helping them prepare activities for their students. These weren't mundane busywork assignments, but, Vera said, "real projects that any astronomer might tackle."[8] Hunter had been running the program for more than a decade, giving students and their teachers an intimate window into what it means to be an astronomer—something Vera would have enjoyed had she had the opportunity as a child.

After Bob's death, this trip would be Vera's last to the telescope. From then on, she analyzed her data at her desk at DTM, still hoping to find more stars far from galaxies' cores and measure if they moved too fast for Newton. And she continued to inspire schoolchildren, early-career scientists, and science writers with her grace, humility, and curiosity. She was always exceptionally generous with the time she gave to the Smithsonian's National Air and Space Museum, and one particular afternoon, she perfectly illustrated the idea that dark matter was necessary in galaxies to make stars move around them the way that her observations showed that they did. The room was full of students from the DC area, and they were there to explore the mysteries of space. The goal of meeting with the students was to give them a better sense of the science Vera would be discussing at an evening lecture.

Vera answered the students' questions, but also gladly participated in a game that might better help the students understand the way a galaxy rotates. She and museum staff took the students into the parking lot beneath the museum, then had them line up at different distances from a central point along circles on the floor. Each student became a star in a galaxy, some closer to the galactic center and some much farther away. The students then started walking at the same pace, simulating stars orbiting the galaxy's center; not long after the students starting walking, Vera told them they were no longer in a line but rather an arc, with the students closer to the galaxy's center ahead of the students farther out. Vera had the students line back up, this time asking them to grab a stiff rod that extended from the arm of the closest student star all the way out to a student star much farther out. The students again started walking and were asked to keep the rod straight. "Soon after, the kids on the outermost circles objected because they had to walk very fast to keep up with the kids on the inner circles," recalled David DeVorkin, a senior curator and historian of astronomy.

When the kids heard Vera speak later that evening, they drew on the experience of being a star in a galaxy and could better relate to the question Vera wanted the audience to consider: Why did the stars farther out in the galaxy move faster than expected? "This exercise illustrated Vera's genius," DeVorkin wrote. "With just a few kids holding hands and a metal rod, Vera not only illustrated how dark matter was discovered, but inspired excitement in the kids."[9]

In her final trips to the telescope, Vera was still searching for those stars— the students in her museum experiment who would be at the very far end of the rod, at the very edges of a galaxy. She wanted to know if those stars moved as fast, working as hard as the students in her simulation did, to keep pace with the stars closer in. And she was looking for stars that didn't keep pace. Finding those stars would reveal to astronomers where a galaxy's cloud of dark matter finally drops off—or if it does at all.

This was important to understand, she said, because if our galaxy's halo reaches farther than scientists think it does, it might be big enough to touch Andromeda. And other galaxies might extend far enough out to tap the halos of their neighbors. Some scientists believed this to be so, and if it were, Vera said, there would be no place in the cosmos without dark matter, at least a little bit of it, making the material not only mysterious but also pervasive. It was an idea Vera clung to during her last observing runs, along with the majesty of observing itself.

"Observing the sky is really thrilling, very very spectacular. Being on a mountain and seeing the sky. . . . People assume that if you do it long enough, you get used to it," she once wrote. "That is not true. There is a curiosity but also a kind of wonder that we are here, attempting to understand it."[10]

Epilogue

Up there in the night sky, light-years from Earth, is a galaxy called ESO 323-G064. It's a diffuse clump of stars, gas, and dust—what astronomers call a low-surface-brightness galaxy—and it's one of the last galaxies Vera wrote about in the astronomical literature. At eighty years old, she refused to give up her work on galaxies or her office at the Department of Terrestrial Magnetism in Washington. She made the final observing run of her career in 2008, when she accompanied Deidre Hunter to Lowell Observatory to take her last look at galaxies and work with local teachers and schoolchildren. A few months after returning from that trip, Vera and a team of four other astronomers published a paper mapping the distribution of dark matter in ESO 323-G064.

The work was important because astronomers had been spending a lot of time probing low-surface-brightness galaxies. In the 1990s and early 2000s, these galaxies took center stage as astronomers pushed to discover how dark matter is spread throughout the clumps of stars, gas, and dust in the universe and throughout the universe itself. Examining low-surface-brightness galaxies closely revealed that they are distinct among galaxy types—they don't have as many stars as the bright, brilliant spirals such as Andromeda. But dim, diffuse low-surface-brightness galaxies can be just as big as or bigger than the brilliant spirals and still have stars circling around them at similar speeds, whether those stars are close in or far out. That suggests that low-surface-brightness galaxies are packed with dark matter, perhaps much more dark matter than the spirals that revealed the mysterious matter's existence. In some cases, astronomers observed, low-surface-brightness galaxies may be made almost entirely of dark matter.

ESO 323-G064 revealed something even more tantalizing: it has a central bulge like regular, bright spiral galaxies, but the disk of stars around that bulge is dim, meaning there are fewer stars there than in regular galaxies. Even still, all the stars move at similar velocities around the galaxy's middle. One way to explain those velocities, Vera and her colleagues suggested, was if there's more dark matter near the core of ESO 323-G064 than what's inferred to be near the cores of bright spirals. If ESO 323-G06 does, in fact, have a lot more dark matter near its center, the discovery could offer astronomers another clue to how the mysterious matter situates itself in galaxies across the cosmos.[1]

Another clue about how dark matter is distributed in the universe comes from other low-surface-brightness galaxies called dwarf galaxies. These have only a quarter to half the stars of the Milky Way—only about 100 billion compared with our galaxy's 250 billion to 400 billion. Despite dwarf galaxies' diminutive size, however, they've played an essential role in the structure and evolution of the universe. They are considered a product of the big bang and the building blocks of the much more massive swirls of stars found in the cosmos. Vera, with Deidre and their collaborators, was still probing the inner workings of these very small galaxies and the very largest ones until her final days at the Department of Terrestrial

Magnetism. In fact, her last two published papers were discussing her hunt for stars very, very far from the cores of NGC 801 and UGC 2885. Those were the two behemoth stellar cities Vera and Deidre were studying at Kitt Peak and Lowell on Vera's final trips to the telescope, where she took her last look at the stars from a telescope's catwalk. She savored every moment, as if she knew it'd be her last there.

But sentiment never got in Vera's way as she pushed the limits of those telescopes, a persistence that paid off. After analyzing the data from the observing runs, Vera and her collaborators reported on hints of significant star formation around 200,000 light-years from each galaxy's core. That's so far from the galactic middle that the gas there is diffuse enough that it doesn't often collapse and condense to form stars. How those stars are born when the gas molecules out there are so far apart is another mystery that Vera left to be solved.[2] She had, as Princeton theoretical astrophysicist Jim Peebles had said, continued to raise more questions than she settled.

Those questions she left are her legacy, as are the cosmic objects that bear her name. Among her many legacies, one of her favorite galaxies, the giant UGC 2885, has been dubbed "Rubin's galaxy." It's one of the largest known spirals in the universe and is also known as the Godzilla galaxy. There were other honors too. Because of her contributions to astronomy and her advocacy for women and minorities in science, Vera was awarded prestigious prizes, including the Bruce Medal of the Astronomical Society of the Pacific, the Peter Gruber International Cosmology Prize, the National Medal of Science, the James Craig Watson Medal, and the Gold Medal of the Royal Astronomical Society. She was only the second woman to receive the Royal Society honor, earning it 168 years after Caroline Herschel claimed the first one for a woman in 1828. Vera was also the second female astronomer elected to the National Academy of Sciences, after her mentor and role model, Margaret Burbidge.

There was one prize Vera did not win, though many people thought she should have: the Nobel. Vera's work, noted Harvard theoretical physicist Lisa Randall, provided convincing evidence of the existence of dark matter. It opened the floodgates for a vast field of follow-on scientific work exploring the nature and structure of the universe. "Of all the great advances in physics during the 20th century, surely this one should rank near the top, making it well deserving of the world's preeminent award in the field."[3]

Critics argued that Vera did not deserve the Nobel Prize because she was not the first astronomer to call attention to flat rotation curves, the prime indirect evidence that dark matter exists. She even credited Mort Roberts and his colleagues for pointing out that galaxies' rotation curves are flat, astrophysicist Scott Tremaine once wrote. "Like Saul after his conversion on the road to Damascus," he explained, "Rubin accepted a revolutionary idea after it was fully formulated, and she became one of its most effective advocates."[4]

Randall concedes there are arguments as to why Vera did not win the award. Maybe the indirect evidence from rapidly rotating galaxies wasn't enough to argue the existence of dark matter. Maybe she wasn't the one responsible for interpreting the meaning of flat rotation curves, and, yes, maybe she was one of many scientists who worked tirelessly to piece together the puzzle of dark matter's existence. But she did convince the astronomy community that flat rotation curves were real, and as John Herschel famously said, "He who proves, discovers."[5]

Another argument used to explain why Vera didn't win a Nobel was that dark matter is still technically theoretical. We don't know what it is, and with some researchers working on alternative theories to explain it away, some critics suggest we wait to dole out the world's most prestigious award until we know the exact nature of dark matter. But that's exactly the opposite of what was done with the discovery of dark energy. The men who led the teams that discovered the accelerated expansion of the universe, an acceleration thought to be driven by dark energy, won the Nobel in 2011—a little more than a decade after the teams had made their discovery—and the award was given even though scientists did not have a clear idea of what dark energy actually is (and still don't). Vera's flat rotation curves, observational data that provided evidence for the existence of dark matter, are analogous to the discovery of the data that showed the universe had an accelerated expansion, indicating the existence of dark energy. Yet Vera went without a Nobel.

Maybe she went without the award because she was a woman; she was overlooked because of her gender, Randall argues, and she wasn't the only woman forgotten when it comes to winning a Nobel Prize. Rosalind Franklin was overlooked for her work on the structure of DNA. So was astronomer Jocelyn Bell, who discovered radio pulsars. Nuclear physicist Chien-Shiung Wu, whose experiments showed that some subatomic particles can tell the difference between left and right, was also forgotten, though two men won the prestigious honor for developing the theory behind her work. Another man later earned a Nobel for a related, seemingly more subtle, discovery. Vera too was neglected, which was a shame, Randall says. "Imagine how many more people she would have reached if her name was also on the list of laureates."[6]

Whether Vera deserved a Nobel prize will probably be debated for decades to come. She'd even said that if she'd won it, she'd be thrilled, but she wasn't sure she really wanted it. It changed your life, she feared, and not always in a good way. "We know a lot of Nobel laureates, and I did see what it did to their lives," Vera's sister said. The prize wasn't what mattered; pursuing science did. And so that's what Vera did for as long as she could. She studied the stars, she mentored scientists, writers, myself included, and she battled daily the heartache that came with the death of her husband in 2008 and then of her daughter in 2014, both events that some would say contributed to her mental decline.

As time passed, she could no longer remember the details of the galaxies she'd explored over the decades or the face of her own sister. The decline became most noticeable in 2015, after Vera's son Allan had moved her from Washington, where she'd lived with Bob since the early 1950s, to Princeton, New Jersey. One day, Ruth called Vera to say she and her husband would be visiting the next day, and Ruth recalled she could hear the excitement in her sister's voice. When Ruth and her husband arrived to visit Vera, they saw her sitting in her down jacket; it was cold.

Ruth said hello.

Vera looked at her and simply said: "Who are you? I'm waiting for my sister."

Now it was Ruth who was heartbroken. Seeing her sister's mental deterioration made her ache inside. "It was a tragedy," she said, her voice cracking as she fought back tears. "If there was anything Vera dreaded, it was losing her mind."[7]

Vera passed away on December 25, 2016.

She left behind a legacy of curiosity and discovery, an endless list of students and colleagues who would say they owe their careers to her, and a recollection of her persistence and grace that continues to inspire us all.

So that we will never forget her, Vera has been memorialized with a ridge on Mars, an asteroid, a galaxy, and most recently a telescope—the first national observatory named for a woman. Fittingly, the observatory headquarters are in Tucson, Arizona, not far from where Vera first observed the night sky as a *real* astronomer. The telescope itself—the Vera C. Rubin Observatory—sits atop Cerro Pachón, a mountain in Chile only ten kilometers south of Cerro Tololo, where Vera spent hours staring into space searching for clues as to what makes galaxies spin. The eye of Vera's observatory is a mammoth 8.4-meter reflecting telescope, which will continue the legacy of the observatory's namesake by scrutinizing wide swaths of the night sky similar to the way Vera surveyed galaxies. The observatory's goal is to address many of the questions that Vera's work inspired: What does our galaxy, the Milky Way, look like? What is dark matter? Where is dark matter? How does dark matter shape the large-scale structure of the cosmos? Beyond those curiosities, the observatory will probe deeper, searching for clues to the nature of dark energy and how it behaves over cosmic time. We may not have to wait long for data directed at those questions to stream in. If all goes well, in a few years, the dome of the Vera Rubin Observatory will trundle open, giving the mammoth telescope inside its first peek at the night sky. As Vera did, it will scan the night sky, determined to resolve the remaining mysteries of the universe.

Acknowledgments

To recognize everyone who made this book possible would be a book itself, so for brevity, I thank Jermey Matthews, who invited me to write this book for the MIT Press. I acknowledge my long-time mentor, Tom Siegfried, for reading and providing feedback on the early drafts of each chapter and continually reminding me I could write this book, even when I doubted myself. David DeVorkin, who first introduced me to Vera Rubin when I was starting as a science writer, also provided invaluable guidance and input. Marcia Bartusiak, my mentor when I was a student at MIT, shaped the early outline of this work, and astronomer Victoria Strait provided feedback on many of the scientific technicalities. I also thank the Rubin family for allowing me to correspond with them throughout the writing process; Vera and Robert Rubin, who spent time with me more than a decade ago patiently answering my questions; and Vera, for allowing me to accompany her while she observed at Kitt Peak National Observatory. Finally, a huge thank you to Dr. Mark Littmann and Bonnie Hufford for introducing me to science writing as a career; and to my family and friends, who have let me share stories about dark matter, astronomy, and so much other incredible science for hours on end.

Notes

Prologue

1. Vera C. Rubin and Deidre Hunter, "A Search for H(Alpha) Emission in the Far Outer Discs of Extremely Large Spiral Galaxies," 2019, NOAO: Proposal Information for 2007B-0170, November 26 (2007), https://www.noao.edu/perl/abstract?2007B-0170.

2. Vera C. Rubin, Norbert Thonnard, and W. Kent Ford Jr., "Rotational Properties of 21 Sc Galaxies with a Large Range of Luminosities and Radii, from Ngc 4605 /R=4kpc/ to Ugc 2885 /R=122 Kpc/," *Astrophysical Journal* 238 (1980): 471.

3. Opening adapted from Ashley Yeager, "Vera Rubin's Universe," *Sky and Telescope* (August 2017).

Chapter 1

1. Ruth Berg, phone interview with the author, August 24, 2020.

2. Vera Rubin, "Session I," interview by David DeVorkin, American Institute of Physics, September 21, 1995, www.aip.org/history-programs/niels-bohr-library/oral-histories/5920-1.

3. Vera C. Rubin, "An Interesting Voyage," *Annual Review of Astronomy and Astrophysics* 49 (2011).

4. Rubin, "Session I."

5. Vera Rubin, "Women's Work," *Science* 86 (1986): 58–65.

6. Renee Bergland, *Maria Mitchell and the Sexing of Science* (Boston: Beacon Press, 2008), 57.

7. "For a Kinder, Cooler America," *Attic*, last modified October 3, 2018, accessed October 10, 2019, https://www.theattic.space/home-page -blogs/2018/10/3/miss-mitchells-comet.

8. Vera Rubin, "Vera Rubin," interview by Alan Lightman, American Institute of Physics, April 3, 1989, www.aip.org/history-programs/niels -bohr-library/oral-histories/33963.

9. Ruth Berg, phone interview with the author, August 24, 2020.

10. Harlow Shapley, "Studies Based on the Colors and Magnitudes in Stellar Clusters. XII. Remarks on the Arrangement of the Sidereal Universe," *Astrophysical Journal* 49 (1919): 311–336.

11. Harlow Shapley, "On the Existence of External Galaxies," *Publications of the Astronomical Society of the Pacific* 31, no. 183 (1919): 261.

Chapter 2

1. Gianfranco Bertone and Dan Hooper, "History of Dark Matter," *Review of Modern Physics* 90 (2018): 045002.

2. Friedrich Wilhelm Bessel, "On the Variations of the Proper Motions of Procyon and Sirius," *Monthly Notices of the Royal Astronomical Society* 6 (August 1844): 136.

3. Davor Krajnović, "The Contrivance of Neptune," *Astronomy and Geophysics* 57, no. 5 (2016): 5.28–5.34, https://doi.org/10.1093/astrogeo /atw183.

4. "Precession of the Perihelion of Mercury," Lawrence Berkeley National Laboratory, accessed December 15, 2019, http://aether.lbl.gov /www/classes/p10/gr/PrecessionperihelionMercury.htm.

5. Urbain Le Verrier, "Lettre de M. Le Verrier à M. Faye sur la théorie de Mercure et sur le mouvement du périhélie de cette planète,"

Comptes rendus hebdomadaires des séances de l'Académie des sciences (Paris) 49 (1859): 379–383.

6. David Malin and Dennis Di Cicco, "Astrophotography—The Amateur Connection, the Roles of Photography in Professional Astronomy, Challenges and Changes" (2010), https://web.archive.org/web/20090110170500/http://encyclopedia.jrank.org/articles/pages/1115/Astrophotography.html

7. Arthur Cowper Ranyard, *Knowledge* 17 (1894); Bertone and Hooper, "History of Dark Matter."

8. Angelo Secchi, *L'Astronomia in Roma nel pontificato di Pio IX: memoria* (Rome: Tipografia della Pace, 1877).

9. Bertone and Hooper, "History of Dark Matter."

10. William Thomson Kelvin, "Lecture XVI," in *Baltimore Lectures on Molecular Dynamics and the Wave Theory of Light*, 260–278 (Cambridge: Cambridge University Press, 2010), https://doi.org/10.1017/CBO9780511694523.020.

11. Henri Poincaré, "The Milky Way and the Theory of Gases," *Popular Astronomy* 14 (1906).

12. Henri Poincaré and Henri Vergne, *Leçons Sur les hypothèses cosmogoniques: Professées à la Sorbonne* (Paris: Librairie Scientifique A. Hermann et Fils, 1911).

13. Ernst Öpik, "Selective Absorption of Light in Space, and the Dynamics of the Universe," *Bulletin de la Société Astronomique de Russie* 21 (1915).

14. Jacobus Cornelius Kapteyn, "First Attempt at a Theory of the Arrangement and Motion of the Sidereal System," *Astrophysical Journal* 55 (1922).

15. James H. Jeans, "The Motions of Stars in a Kapteyn Universe," *Monthly Notices of the Royal Astronomical Society* 82 (1922).

16. Virginia Trimble, "Existence and Nature of Dark Matter in the Universe," *Annual Review of Astronomy and Astrophysics* 25 (1987).

17. Jan Hendrik Oort, "The Force Exerted by the Stellar System in the Direction Perpendicular to the Galactic Plane and Some Related Problems," *Bulletin of the Astronomical Institutes of the Netherlands* 6 (1932).

18. Fritz Zwicky, "Die Rotverschiebung von extragalaktischen Nebeln," *Helvetica Physica Acta* 6 (1933). Bertone and Hooper, "History of Dark Matter."

19. Sinclair Smith, "The Mass of the Virgo Cluster," *Astrophysical Journal* 83 (1936).

20. Edwin P. Hubble, *Realm of the Nebulae* (New Haven: Yale University Press 1936).

21. Fritz Zwicky, "On the Masses of Nebulae and of Clusters of Nebulae," *Astrophysical Journal* 86 (1937).

22. Horace W. Babcock, "Spectrographic Observations of the Rotation of the Andromeda Nebula," *Publications of the Astronomical Society of the Pacific*, 50 no. 295 (1938): 174; Horace W. Babcock, "The Rotation of the Andromeda Nebula," *Lick Observatory Bulletins* 498 (Berkeley: University of California Press 1939): 41–51.

23. Erik Holmberg, "On the Clustering Tendencies among the Nebulae," *Astrophysical Journal* 92 (1940).

Chapter 3

1. Vera Rubin, "Session I," interview by David DeVorkin, American Institute of Physics, September 21, 1995, www.aip.org/history-programs /niels-bohr-library/oral-histories/5920-1.

2. Vera Rubin, "Vera Rubin," interview by Alan Lightman, American Institute of Physics, April 3, 1989, www.aip.org/history-programs/niels -bohr-library/oral-histories/33963.

3. Rubin, "Session I."

4. Rubin, "Session I."

5. Rubin, "Session I."

6. Rubin, "Session I."

7. "Woman Hunts Waifs of Sky," *Oakland Tribune*, September 9, 1928, https://www.newspapers.com/clip/3019405/maud_w_makemson _as_a_doctoral_student/.

8. Rubin, "Session I."

9. Vera C. Rubin, "An Interesting Voyage," *Annual Review of Astronomy and Astrophysics* 49 (2011).

10. Rubin, "Session I."

11. Rubin, "Session I."

12. Rubin, "Session I."

13. Vera Rubin, "Vera Rubin," interview by David DeVorkin and Ashley Yeager, American Institute of Physics, July 20, 2007, https://www.aip .org/history-programs/niels-bohr-library/oral-histories/44082.

14. Rubin, "Session I."

Chapter 4

1. Vera Rubin, "Session I," interview by David DeVorkin, American Institute of Physics, September 21, 1995, www.aip.org/history-programs /niels-bohr-library/oral-histories/5920-1.

2. Rubin, "Session I," 1995.

3. Rubin, "Session I," 1995.

4. O. Neugebauer, "The History of Ancient Astronomy Problems and Methods," *Journal of Near Eastern Studies* 4 (1945).

5. Steven Weinberg, *The First Three Minutes* (New York: Basic Books, 1977), 128–129.

6. Ralph Alpher, Hans Bethe, and George Gamow, "The Origin of Chemical Elements," *Physical Review* 73 (1948); Ralph Alpher, "A Neutron-Capture Theory of the Formation and Relative Abundance of the Elements," *Physical Review* 74 (1948); Ralph Alpher, "Origin and

Relative Abundance of the Chemical Elements" (George Washington University, 1948).

7. George Gamow, *The Creation of the Universe* (New York: Viking Press, 1952).

8. Physics CU Boulder, "George Gamow, Gifted Physicist," YouTube, April 1, 2015, https://www.youtube.com/watch?v=Y3wNPzuJuwc.

9. George Gamow and Edward Teller, "The Expanding Universe and the Origin of the Great Nebulae," *Nature* 143 (1939); George Gamow and Edward Teller, "On the Origin of Great Nebulae," *Physical Review* 55 (1939).

10. George Gamow, "Rotating Universe?" *Nature* 158 (1946).

11. Grote Reber, "Notes: Cosmic Static," *Astrophysical Journal* 91 (1940).

12. Jan Oort, "Jan Oort," interview by David DeVorkin, American Institute of Physics, November 10, 1977, https://www.aip.org/history -programs/niels-bohr-library/oral-histories/4806.

13. Vera Rubin, "Vera Rubin," interview by David DeVorkin and Ashley Yeager, American Institute of Physics, July 20, 2007, https://www.aip .org/history-programs/niels-bohr-library/oral-histories/44082.

14. Vera Rubin, "Vera Rubin," interview by Alan Lightman, American Institute of Physics, April 3, 1989, www.aip.org/history-programs /niels-bohr-library/oral-histories/33963.

Chapter 5

1. Vera Rubin, "Session II," interview by David DeVorkin, American Institute of Physics, May 9, 1996, www.aip.org/history-programs/niels -bohr-library/oral-histories/5920-2.

2. Vera C. Rubin, "An Interesting Voyage," *Annual Review of Astronomy and Astrophysics* 49 (2011).

3. Kristine Larsen, "Reminiscences on the Career of Martha Stahr Carpenter: Between a Rock and (Several) Hard Places," *JAAVSO* 40 (2012), http://www.aavso.org/sites/default/files/jaavso/v40n1/51.pdf;

J. H. Moore, "Survey of the Year's Work at the Lick Observatory," *Publications of the Astronomical Society of the Pacific* 58, no. 340 (1946).

4. George Gamow, "Rotating Universe?" *Nature* 158 (1946).

5. Vera Rubin, "Vera Rubin," interview by Alan Lightman, American Institute of Physics, April 3, 1989, www.aip.org/history-programs/niels-bohr-library/oral-histories/33963.

6. Rubin, "Session II."

7. Rubin, "Vera Rubin," 1989.

8. Rubin, "Session II."

9. Rubin, "Vera Rubin," 1989.

10. Howard Blakeslee, "Student Says Stars May Show Creation's Center," *Ithaca Journal*, December 30, 1950.

11. Vera C. Rubin, "Differential Rotation of the Inner Metagalaxy," *Astronomical Journal* 56 (1951).

12. Rubin, "An Interesting Voyage."

13. Rubin, "Session II."

14. Rubin, "Session II."

15. Ralph Alpher, Hans Bethe, and George Gamow, "The Origin of Chemical Elements," *Physical Review* 73 (1948); Ralph Alpher, "A Neutron-Capture Theory of the Formation and Relative Abundance of the Elements," *Physical Review* 74 (1948); Ralph A. Alpher, "Origin and Relative Abundance of the Chemical Elements" (George Washington University, 1948).

16. Rubin, "Session II."

17. Vera Rubin, "Vera Rubin," interview by David DeVorkin and Ashley Yeager, American Institute of Physics, July 20, 2007, https://www.aip.org/history-programs/niels-bohr-library/oral-histories/44082.

18. Rubin, "Session II."

19. Rubin, "Session II."

Chapter 6

1. Vera Rubin, "Session II," interview by David DeVorkin, American Institute of Physics, May 9, 1996, www.aip.org/history-programs/niels-bohr-library/oral-histories/5920-2.

2. Vera Rubin, "George Gamow," in *Bright Galaxies, Dark Matters* (Woodbury, NY: American Institute of Physics, 2007) 187.

3. Rubin, "Session II."

4. Martin Schwarzschild, letter to Vera Rubin, August 20, 1952, Vera C. Rubin Papers, Manuscript Division, Library of Congress, Washington, DC, Box 29.

5. Email correspondence from Vera Rubin to Judith Rubin, September 15, 1992.

6. Rubin, "Vera Rubin—Session II."

7. Vera Rubin, "Vera Rubin," interview by Alan Lightman, American Institute of Physics, April 3, 1989, www.aip.org/history-programs/niels-bohr-library/oral-histories/33963.

8. Rubin, "Session II."

9. Gerard de Vaucouleurs, "Evidence for a Local Supergalaxy," *Astronomical Journal* 58 (1953).

10. Rubin, "Session II."

11. Fred Hoyle, "The Synthesis of the Elements from Hydrogen," *Monthly Notices of the Royal Astronomical Society* 106 (1946).

12. Rubin, "Session II."

13. Vera C. Rubin, "An Interesting Voyage," *Annual Review of Astronomy and Astrophysics* 49 (2011).

14. Rubin, "Session II." See also David DeVorkin, "The Changing Place of Red Giant Stars in the Evolutionary Process," *Journal for the History of Astronomy* 37 (2006): 429–469.

15. Vera C. Rubin, "Fluctuations in the Space Distribution of the Galaxies," *Proceedings of the National Academy of Sciences* 40 (1954).

Chapter 7

1. Vera Rubin, "Session II," interview by David DeVorkin, American Institute of Physics, May 9, 1996, www.aip.org/history-programs/niels -bohr-library/oral-histories/5920-2.

2. Jan H. Oort, "The Structure of the Cloud of Comets Surrounding the Solar System and a Hypothesis Concerning Its Origin," *Bulletin of the Astronomical Institutes of the Netherlands* 11 (1950).

3. Hugo van Woerden and Richard G. Strom, "The Beginnings of Radio Astronomy in the Netherlands," *Journal of Astronomical History and Heritage* 9 (2006).

4. J. Oort (in collaboration with C. A Muller), "Spiral Structure and Interstellar Emission," *Monthly Notes of the Astronomical Society of South Africa* 11 (1952).

5. Van Woerden and Strom, "The Beginnings of Radio Astronomy."

6. Vera C. Rubin, "The Form of the Galactic Spiral Arms from a Modified Oort Theory," *Astronomical Journal* 60 (1955).

7. Vera Rubin, "Vera Rubin," interview by David DeVorkin and Ashley Yeager, American Institute of Physics, July 20, 2007, https:// www.aip.org/history-programs/niels-bohr-library/oral-histories /44082.

8. E. Margaret Burbidge, "E. Margaret Burbidge," interview by David DeVorkin, American Institute of Physics, July 13, 1978, www.aip.org /history-programs/niels-bohr-library/oral-histories/25487.

9. Vera C. Rubin, "E. Margaret Burbidge," in *Bright Galaxies, Dark Matters* (Woodbury, NY: American Institute of Physics, 2007), 191.

10. Fred Hoyle, William A. Fowler, Geoffrey R. Burbidge, and E. Margaret Burbidge, "Origin of the Elements in Stars," *Science* 124 (1956).

11. E. Margaret Burbidge and Geoffrey R. Burbidge, "Rotation and Internal Motions in NGC 5128," *Astrophysical Journal* 129 (1959).

12. Rubin, "Vera Rubin."

13. Maud Makemson, letter to Vera Rubin, 1957, Vera C. Rubin Papers, Manuscript Division, Library of Congress, Washington, DC, Box 29.

14. Vera C. Rubin, "Solar Limb Darkening Determined from Eclipse Observations," *Astrophysical Journal* 129 (1959).

15. Vera C. Rubin, "An Interesting Voyage," *Annual Review of Astronomy and Astrophysics* 49 (2011).

16. Email from Allan Rubin to the author, August 18, 2020.

17. Rubin, "An Interesting Voyage."

18. Vera C. Rubin, "Evolution of the Galactic System," *Physics Today* 13 (1960).

19. Vera Rubin, letter to Gérard de Vaucouleurs, November 12, 1960, Vera C. Rubin Papers, Manuscript Division, Library of Congress, Washington, DC.

20. V. A. Ambartsumian, "Multiple Systems of Trapezium type," *Soobshcheniya Byurakanskoj Observatorii Akademiya Nauk Armyanskoj SSR Erevan* 15 (1954).

21. Jim Peebles, *Cosmology's Century* (Princeton, NJ: Princeton University Press, 2020).

22. Jerzy Neyman, Thornton Page, and Elizabeth Scott, "Conference on the Instability of Systems of Galaxies," *Astronomical Journal* 66 (1961).

23. Neyman et al., "Conference on the Instability of Systems of Galaxies."

Chapter 8

1. Vera C. Rubin, J. Burley, A. Kiasatpoor, B. Klock, B. G. Pease, E. Rutscheidt, and C. Smith, "Comparison of Radio and Optical Radial

Velocity Data in the Vicinity of the Sun," *Astronomical Journal* 67 (1962).

2. Vera Rubin, J. Burley, A. Kiasatpoor, B Klock, G. Pease, E. Rutscheidt, and C. Smith, "Kinematic Studies of Early-Type Stars. I. Photometric Survey, Space Motions, and Comparison with Radio Observations," *Astronomical Journal* 67 (1962).

3. Vera C. Rubin, "An Interesting Voyage," *Annual Review of Astronomy and Astrophysics* 49 (2011).

4. S. Stephens, "An Unconventional Career," *Mercury* 21 (1992).

5. Rubin, "An Interesting Voyage."

6. Allan Sandage, "Current Problems in the Extragalactic Distance Scale," *Astrophysical Journal* 127 (1958).

7. Vera Rubin, "Vera Rubin," interview by David DeVorkin and Ashley Yeager, American Institute of Physics, July 20, 2007, https://www.aip .org/history-programs/niels-bohr-library/oral-histories/44082.

8. Rubin, "An Interesting Voyage."

9. Rubin, "Vera Rubin."

10. David S. Evans and J. Derral Mulholland, *Big and Bright: A History of the McDonald Observatory* (Austin: University of Texas Press, 2013).

11. Rubin, "An Interesting Voyage."

Chapter 9

1. Vera Rubin, interview by the author, Kitt Peak National Observatory, November 12–14, 2007.

2. Vera C. Rubin, "An Interesting Voyage," *Annual Review of Astronomy and Astrophysics* 49 (2011).

3. Vera Rubin, "Vera Rubin," interview by David DeVorkin and Ashley Yeager, American Institute of Physics, July 20, 2007, https://www.aip .org/history-programs/niels-bohr-library/oral-histories/44082.

4. Rubin, "Vera Rubin."

5. Vera C. Rubin, E. Margaret Burbidge, Geoffrey R. Burbidge, D. J. Crampin, and Kevin H. Prendergast, "The Rotation and Mass of NGC 7331," *Astrophysical Journal* 141 (1964).

6. Kent Ford, telephone interview by Ashley Yeager, Cambridge, MA, October 31, 2007.

7. Email from Allan Rubin to the author, November 18, 2019.

8. Ford interview.

9. Ford interview.

10. Vera Rubin, "Session II," interview by David DeVorkin, American Institute of Physics, May 9, 1996, www.aip.org/history-programs/niels -bohr-library/oral-histories/5920-2.

11. Rubin, "An Interesting Voyage."

12. Vera C. Rubin and Kent W. Ford, "Image Tube Spectra of Quasi-Stellar Objects," *Astronomical Journal* 71 (1966).

13. Kent Ford, "W. Kent Ford, Jr.," interview by David DeVorkin and Shaun Hardy, American Institute of Physics, October 25, 2013, www .aip.org/history-programs/niels-bohr-library/oral-histories/43241.

14. Rubin, "An Interesting Voyage."

15. Vera Rubin, interview by the author, Kitt Peak National Observatory, November 12–14, 2007.

Chapter 10

1. Morton S. Roberts, "A High-Resolution 21-CM Hydrogen-Line Survey of the Andromeda Nebula," *Astrophysical Journal* 144 (1966): 639.

2. Vera Rubin, interview by the author, Kitt Peak National Observatory, November 12–14, 2007.

3. Vera Rubin, "Vera Rubin," interview by David DeVorkin and Ashley Yeager, American Institute of Physics, July 20, 2007, https://www.aip .org/history-programs/niels-bohr-library/oral-histories/44082.

4. Horace W. Babcock, "Spectrographic Observations of the Rotation of the Andromeda Nebula," *Publications of the Astronomical Society of the Pacific* 50, no. 295 (1938): 174; Horace W. Babcock, "The Rotation of the Andromeda Nebula," *Lick Observatory Bulletins* 498 (1939): 41–51.

5. Rubin interview by Yeager.

6. Vera C. Rubin, "An Interesting Voyage," *Annual Review of Astronomy and Astrophysics* 49 (2011): 1–28.

7. Rubin, "An Interesting Voyage."

8. Rubin, "An Interesting Voyage."

9. Rubin interview by Yeager.

10. Email from Allan Rubin to the author, November 2019.

11. Vera C. Rubin and Kent W. Ford Jr., "Rotation of the Andromeda Nebula from a Spectroscopic Survey of Emission Regions," *Astrophysical Journal* 159 (1970): 379.

12. Ken C. Freeman, "On the Disks of Spiral and S0 Galaxies," *Astrophysical Journal* 160 (1970): 811.

13. "How Dark Matter Became a Particle," *CERN Courier* 57:4 (2017) : 26–33, https://cds.cern.ch/record/2265254

14. Kent W. Ford Jr., Vera C. Rubin, and Morton S. Roberts, "A Comparison of 21-cm Radial Velocities and Optical Radial Velocities of Galaxies," *Astronomical Journal* 76 (1971): 22–24.

15. Sandra Faber, "Sandra Faber," interview by Alan Lightman, American Institute of Physics, October 15, 1988, www.aip.org/history -programs/niels-bohr-library/oral-histories/33932

16. Faber, "Sandra Faber."

Chapter 11

1. Marcia Bartusiak, *Black Hole* (New Haven, CT: Yale University Press, 2015).

2. Seth Shostak, "Aperture Synthesis Observations of Neutral Hydrogen in Three Galaxies" (PhD diss., California Institute of Technology, 1972).

3. David H. Rogstad and Seth G. Shostak, "Gross Properties of Five Scd Galaxies as Determined from 21-Centimeter Observations," *Astrophysical Journal* 176 (1972): 315.

4. Seth Shostak, phone interview with the author, March 10, 2020.

5. Vera C. Rubin and John M. Losee, "A Finding List of Faint Blue Stars in the Anticenter Region of the Galaxy," *Astronomical Journal* 76 (1971): 1099–1101.

6. Morton S. Roberts and Arnold H. Rots, "Comparison of Rotation Curves of Different Galaxy Types," *Astronomy and Astrophysics* 26 (1973): 483–485.

7. Alar Toomre, "On the Gravitational Stability of a Disk of Stars," *Astrophysical Journal* 139 (1964): 1217–1238.

8. Frank Hohl, "Dynamical Evolution of Disk Galaxies," NASA Tech. Rep., NASA-TR R-343 (1970).

9. Richard H. Miller, Kevin H. Prendergast, and William J. Quirk, "Numerical Experiments on Spiral Structure," *Astrophysical Journal* 161 (1970): 903–916.

10. Hohl, "Dynamical Evolution of Disk Galaxies," 1970.

11. P. J. E. Peebles, *Cosmology's Century: An Inside History of Our Modern Understanding of the Universe* (Princeton, NJ: Princeton University Press, 2020).

12. A. Penzias and R. W. Wilson, "A Measurement of Excess Antenna Temperature At 4080 Mc/s," *Astrophysical Journal Letters* 142 (1965); R. H. Dicke, P. J. E. Peebles, P. J., Roll, and D. T. Wilkinson, "Cosmic Black-Body Radiation," *Astrophysical Journal Letters* 142 (1965).

13. Peebles, *Cosmology's Century*.

14. P. J. E. Peebles, "Structure of the Coma Cluster of Galaxies," *Astronomical Journal* 75 (1970): 13.

15. Jeremiah P. Ostriker and P. J. E. Peebles, "A Numerical Study of the Stability of Flattened Galaxies: or, Can Cold Galaxies Survive?" *Astrophysical Journal* 186 (1973): 467–480.

16. Peebles, *Cosmology's Century*.

17. Jaan Einasto, Ants Kaasik, and Enn Saar, "Dynamic Evidence on Massive Coronas of Galaxies," *Nature* 250, (1974): 309–310.

18. Jeremiah Ostriker, phone interview with the author, March 13, 2020.

19. Carl Sagan, "Encyclopaedia Galactica," *Cosmos: A Personal Voyage*, episode 12, aired December 14, 1980, on PBS.

20. Jeremiah P. Ostriker, P. J. E. Peebles, and Amos Yahil, "The Size and Mass of Galaxies, and the Mass of the Universe," *Astrophysical Journal* 193 (1974): L1.

Chapter 12

1. Norbert Thonnard to Vera Rubin, postcard, December 8, 1980, Vera C. Rubin Papers, Manuscript Division, Library of Congress, Washington, DC.

2. W. L. Peters III, "Models for the Inner Regions of the Galaxy. I. An Elliptical Streamline Model," *Astrophysical Journal* 195 (1975), 617–629.

3. C. J. Peterson, V. C. Rubin, W. K. Ford Jr., and N. Thonnard, "Motions of the Stars and Excited Gas in the Barred Spiral Galaxy 3351," *Bulletin of the American Astronomical Society* 7 (1976): 538.

4. Vera C. Rubin and Robert J. Rubin, "Early Observations of the Crab Nebula as a Nebula," *Bulletin of the American Astronomical Society* 5 (1973): 411.

5. Vera C. Rubin, Kent Ford, and Judith Rubin, "A Curious Distribution of Radial Velocities of SCi Galaxies with $14.0 <= M <= 15.0$," *Astrophysical Journal* 183 (1973): L111.

6. Laurent Nottale and Hiroshi Karoji, "Possible Implications of the Rubin-Ford Effect," *Nature* 31–33 (1976).

7. S. M. Fall and B. J. T. Jones, "Isotropic Cosmic Expansion and the Rubin–Ford Effect," *Nature* 262 (1976): 457–460.

8. Rubin et al., "A Curious Distribution of Radial Velocities."

9. Jim Peebles, interview by Christopher Smeenk, April 4, 2002, Niels Bohr Library and Archives, American Institute of Physics, College Park, MD, http://www.aip.org/history-programs/niels-bohr-library/oral-histories/25507-1.

10. Peebles interview.

11. D. L. Hawley and P. J. E. Peebles, "Distribution of Observed Orientations of Galaxies," *Astronomical Journal* 80 (1975): 477–491.

12. P. J. E. Peebles, "The Peculiar Velocity Field in the Local Supercluster," *Astrophysical Journal* 205 (1976): 318–328.

13. V. C. Rubin, W. Kent Ford Jr., Norbert Thonnard, Morton S. Roberts, and John A. Graham, "Motion of the Galaxy and the Local Group Determined from the Velocity Anisotropy of Distant SC I Galaxies. I. The Data," *Astronomical Journal* 81 (1976): 687–718; V. C. Rubin, W. Kent Ford Jr., Norbert Thonnard, Morton S. Roberts, and John A. Graham, "Motion of the Galaxy and the Local Group Determined from the Velocity Anisotropy of Distant SC I Galaxies. II. The Analysis for the Motion," *Astronomical Journal* 81 (1976): 719–737.

14. "Rubin-Ford Effect," in *A Dictionary of Astronomy*, 2nd rev. ed., ed. Ian Redpath (Oxford: Oxford University Press, 2016), 406.

15. Martin Clutton-Brock and Phillip James Edwin Peebles, "Galaxy Clustering and the Rubin-Ford Effect," *Astronomical Journal* 86 (1981): 1115–1119.

16. Clutton-Brock and Peebles, "Galaxy Clustering and the Rubin-Ford Effect."

17. Gregory Bothun, *Modern Cosmological Observations and Problems* (Boca Raton, FL: CRC Press, 1998), 118.

18. D. T. Emerson and J. E. Baldwin, "The Rotation Curve and Mass Distribution in M31," *Monthly Notices of the Royal Astronomical Society* 165, no. 1 (1973): 9P–13P.

19. G. Burbidge, "On the Masses and Relative Velocities of Galaxies," *Astrophysical Journal* 196 (1975): L7–L10.

20. V. C. Rubin, C. J. Peterson, and W. K. Ford Jr., "The Rotation Curve of the E7/SO Galaxy NGC 3115," *Bulletin of the American Astronomical Society* 8 (1976): 297.

21. Jim Peebles, "Vera's Challenge to Modern Cosmology," Rubin Symposium, June 24, 2019.

22. V. C. Rubin, W. K. Ford Jr., C. J. Peterson, and J. H. Oort, "New Observations of the NGC 1275 Phenomenon," *Astrophysical Journal* 211 (1977): 693–696.

23. Norbert Thonnard, phone interview by the author, January 18, 2008.

24. V. C. Rubin, W. K. Ford Jr., and N. Thonnard, "Extended Rotation Curves of High-Luminosity Spiral Galaxies. IV. Systematic Dynamical Properties, Sa → Sc," *Astrophysical Journal* 225 (1978): L107–L111.

25. S. M. Faber and J. S. Gallagher, "Masses and Mass-to-Light Ratios of Galaxies," *Annual Review of Astronomy and Astrophysics* 17 (1979): 135–187.

26. Thomas S. Kuhn, *The Structure of Scientific Revolutions* (Chicago: University of Chicago Press, 1970), 54.

27. Vera Rubin, interview by Ashley Yeager, Kitt Peak National Observatory, November 12–14, 2007. V. C. Rubin, W. K. Ford Jr., and N. Thonnard, "Rotational Properties of 21 SC Galaxies with a Large Range of Luminosities and Radii, from NGC 4605 (R=4kpc) to UGC 2885 (R=122kpc)," *Astrophysical Journal* 238 (1980): 471–487.

28. Jim Peebles, "Vera's Challenge to Modern Cosmology," June 2019.

29. Norbert Thonnard to Rubin, postcard.

30. Vera Rubin, "Vera Rubin," interview by David DeVorkin and Ashley Yeager, American Institute of Physics, July 20, 2007, https://www.aip.org/history-programs/niels-bohr-library/oral-histories/44082.

Chapter 13

1. Vera C. Rubin, "Dark Matter in the Universe," in *Highlights of Astronomy*, vol. 7: *Proceedings of the Nineteenth IAU General Assembly* (Dordrecht: D. Reidel, 1986).

2. François Schweizer, Bradley Whitmore, and Vera C. Rubin, "Colliding and Merging Galaxies. II. SO Galaxies with Polar Rings," *Astronomical Journal* 88 (1983).

3. Woodruff T. Sullivan III, *Cosmic Noise: A History of Early Radio Astronomy* (Cambridge: Cambridge University Press, 2009).

4. Jeremiah Ostriker, phone interview by the author, March 13, 2020.

5. Vera Rubin, "Stars, Galaxies, Cosmos: The Past Decade, the Next Decade," *Science* 209 (1980).

6. Morton S. Roberts and Robert N. Whitehurst, "The Rotation Curve and Geometry of M31 at Large Galactocentric Distances," *Astrophysical Journal* 201 (1975). S. D. M. White and M. J. Rees, "Core Condensation in Heavy Halos: A Two-Stage Theory for Galaxy Formation and Clustering," *Monthly Notices of the Royal Astronomical Society* 183 (1978).

7. P. J. E. Peebles, *Cosmology's Century* (Princeton, NJ: Princeton University Press, 2020).

8. Douglas N. C. Lin and Sandra M. Faber, "Some Implications of Nonluminous Matter in Dwarf Spheroidal Galaxies," *Astrophysical Journal* 266 (1983): L21–L25.

9. Peebles, *Cosmology's Century*.

10. James Gunn, Ben W. Lee, Ian Lerche, David N. Schramm, Gary Steigman, "Some Astrophysical Consequences of the Existence of a Heavy Stable Neutral Lepton," *Astrophysical Journal* 223 (1978).

11. Frank Wilczek, "The Birth of Axions," *Current Contents* 16 (1991).

12. James Ipser and Pierre Sikivie, "Can Galactic Halos Be Made of Axions?" *Physical Review Letters* 50 (1983).

13. Mordehai Milgrom, "A Modification of the Newtonian Dynamics as a Possible Alternative to the Hidden Mass Hypothesis," *Astrophysical Journal* 270 (1983).

14. Rubin, "Dark Matter in the Universe."

Chapter 14

1. Maiken Scott, "Vera Rubin's Son Reflects on How She Paved the Way for Women," *WHYY*, January 12, 2017.

2. "An Unconventional Career," in *Bright Galaxies, Dark Matters* (New York: American Institute of Physics and Springer, 1997) 153–163.

3. Anne Cowley et al., "Report to the Council of the AAS from the Working Group on the Status of Women in Astronomy," *American Astronomical Society Bulletin* 6, no. 3, pt. II (1974): 412–423.

4. Cowley et al., "Report to the Council of the AAS."

5. Ben Skuse, "Celebrating Astronomer Margaret Burbidge, 1919–2020," *Sky and Telescope*, April 6, 2020, https://skyandtelescope.org/astronomy-news/happy-birthday-margaret-burbidge/

6. Stephens, "An Unconventional Career."

7. Vera Rubin, "Sexism in Science," *Physics Today* 31 (1978): 13.

8. Vera C. Rubin, "Male World of Physics?" *Physics Today* 35, no. 5 (1982): 121.

9. Ruth Berg, phone interview with the author, August 24, 2020.

10. Rubin, *Bright Galaxies, Dark Matters*, 172.

11. R. Giovanelli, M. P. Haynes, V. C. Rubin, and W. K. Ford Jr., "UGC 12591: The Most Rapidly Rotating Disk Galaxy," *Astrophysical Journal Letters* 301 (1986): L7.

12. Deidre Hunter, Vera C. Rubin, and John S. Gallagher, "Optical Rotation Velocities and Images of the Spiral Galaxy NGC 3198," *Astronomical Journal* 91 (1986): 1086–1090.

13. D. Burstein, V. C. Rubin, W. K. Ford Jr., and B. C. Whitmore, "Is the Distribution of Mass within Spiral Galaxies a Function of Galaxy Environment?" *Astrophysical Journal Letters* 305 (1986): L11.

14. Vera Rubin, "Coherent Large Scale Motions from a New Sample of Spiral Galaxies," in *Large Scale Structures of the Universe: Proceedings of the 130th Symposium of the International Astronomical Union*, ed. Jean Audouze, Marie-Christine Pelletan, and Sandor Szalay (Dordrecht: Kluwer, 1988), 181.

15. Walter Sullivan, "New View of Universe Shows Sea of Bubbles to Which Stars Cling," *New York Times*, January 5, 1986.

16. Vera C. Rubin, "The Local Supercluster and Anisotropy of the Redshifts," in: Corwin H.G., Bottinelli L. (eds) *The World of Galaxies* (New York: Springer-Verlag, 1989), 431–451, 452.

17. Damond Benningfield, "Vera Rubin," *StarDate* (September/October 1989).

18. Allan Rubin, email to the author, June 24, 2020.

19. Rubin, *Bright Galaxies, Dark Matters*, 173.

20. Rubin, *Bright Galaxies, Dark Matters*, 173.

21. Vera C. Rubin, J. A. Graham, Jeffrey D. P. Kenney, "Cospatial Counterrotating Stellar Disks in the Virgo E7/S0 Galaxy NGC 4550," *Astrophysical Journal Letters* 394 (1992): L9.

22. "National Science Foundation—Where Discoveries Begin." National Medal of Science 50th Anniversary, National Science Foundation, accessed June 1, 2020, at www.nsf.gov/news/special_reports/medalof science50/rubin.jsp.

23. Rubin, *Bright Galaxies, Dark Matters*, xii.

24. Adam Riess et al., "Observational Evidence from Supernovae for an Accelerating Universe and a Cosmological Constant," *Astronomical Journal* 116, no. 3 (1998): 1009–1038; Saul Perlmutter et al., "Measurements of Omega and Lambda from 42 High Redshift Supernovae," *Astrophysical Journal* 517, no. 2 (1999): 565–586.

25. S. Perlmutter, M. Turner, and M. White, "Constraining Dark Energy with Type Ia Supernovae and Large-Scale Structure," *Physical Review Letters* 83, no. 4 (1999): 670–673.

26. George Gamow, *My World Line: An Informal Autobiography* (New York: Viking Press, 1970), 44.

Chapter 15

1. Vera Rubin, interview by the author at Kitt Peak National Observatory, November 12–14, 2007.

2. Rubin interview.

3. Rubin interview.

4. Rubin interview.

5. Chandra X-ray Observatory, "NASA Finds Direct Proof of Dark Matter," Harvard-Smithsonian Center for Astrophysics, August 21, 2006, https://www.chandra.harvard.edu/photo/2006/1e0657/.

6. Patricia Sullivan, "Robert J. Rubin, 81; Scientist Whose Work Combined Disciplines," *Washington Post*, February 5, 2008.

7. Ruth Berg, phone interview with the author, August 24, 2020.

8. Vera C. Rubin, "An Interesting Voyage," *Annual Review of Astronomy and Astrophysics* 49 (2011): 1–28.

9. David DeVorkin, "Capturing the Essence of Astronomer Vera Rubin," Smithsonian Air and Space Museum, December 30, 2016, https://airandspace.si.edu/stories/editorial/capturing-essence-astronomer-vera-rubin

10. Vera Rubin, letter to David Andrews, n.d., Vera C. Rubin Papers, Manuscript Division, Library of Congress, Washington, DC.

Epilogue

1. Lodovico Coccato et al., "VIMOS-VLT Integral Field Kinematics of the Giant Low Surface Brightness Galaxy ESO 323-G064," *Astronomy and Astrophysics* 490, no. 2 (2008) 589–600.

2. Allison Ashburn, Deidre A. Hunter, and Vera C. Rubin, "Star Formation in the Extreme Outer Disks of Giant Spiral Galaxies," American Astronomical Society Meeting 22, id.146.05 (2013).

3. Lisa Randall, "Why Vera Rubin Deserved a Nobel," *New York Times*, January 4, 2017.

4. Scott Tremaine, "Explaining a Few Discoveries," *Physics Today* 70, no. 9 (2017): 12.

5. David Gooding, "'He Who Proves, Discovers': John Herschel, William Pepys and the Faraday Effect," *Notes and Records of the Royal Society of London* 39, no. 2 (1985): 229–244.

6. Randall, "Why Vera Rubin Deserved a Nobel."

7. Ruth Berg, phone interview with the author, August 24, 2020.

Index

Page numbers in italic type indicate figures.